T0212056

Mapping the Moral Geographies of Education

This book explores the growth of 'character education' in schools and youth organisations over the last decade. It delves into historical and contemporary debates through a geopolitical lens.

With a renewed focus on values and virtues such as grit, gumption, perseverance, resilience, generosity and neighbourliness, this book charts the re-imagining and re-fashioning of a 'character agenda' in England and examines its multi-scalar geographies. It explores how these moral geographies of education for children and young people have developed over time. Drawing on original research and examples from schools, military and uniformed youth organisations, and the state-led National Citizen Service, the book critically examines the wider implications of the 'character agenda' across the United Kingdom and beyond. It does so by raising a series of questions about the interconnections between character, citizenship and values and highlighting how these moral geographies reach far beyond the classroom or campsite.

Offering critical insights on the roles of character, citizenship and values in modern education, this book will be of immense value to educationists, teachers and policymakers. It will appeal to students and scholars of human geography, sociology, education studies, cultural studies and history.

Sarah Mills is a Reader in Human Geography at Loughborough University, UK. Her research focuses on the geographies of youth citizenship, informal education and volunteering across contemporary and historical contexts. In 2017, she received the Gill Memorial Award from the Royal Geographical Society with the Institute of British Geographers.

Routledge Spaces of Childhood and Youth Series
Edited by Peter Kraftl and John Horton

The *Routledge Spaces of Childhood and Youth Series* provides a forum for original, interdisciplinary and cutting edge research to explore the lives of children and young people across the social sciences and humanities. Reflecting contemporary interest in spatial processes and metaphors across several disciplines, titles within the series explore a range of ways in which concepts such as space, place, spatiality, geographical scale, movement/mobilities, networks and flows may be deployed in childhood and youth scholarship. This series provides a forum for new theoretical, empirical and methodological perspectives and ground-breaking research that reflects the wealth of research currently being undertaken. Proposals that are cross-disciplinary, comparative and/or use mixed or creative methods are particularly welcomed, as are proposals that offer critical perspectives on the role of spatial theory in understanding children and young people's lives. The series is aimed at upper-level undergraduates, research students and academics, appealing to geographers as well as the broader social sciences, arts and humanities.

After Childhood
Re-thinking Environment, Materiality and Media in Children's Lives
Peter Kraftl

Why Garden in Schools?
Lexi Earl and Pat Thomson

Latin American Transnational Children and Youth
Experiences of Nature and Place, Culture and Care Across the Americas
Edited by Victoria Derr and Yolanda Corona-Caraveo

Mapping the Moral Geographies of Education
Character, Citizenship and Values
Sarah Mills

For more information about this series, please visit: https://www.routledge.com/Routledge-Spaces-of-Childhood-and-Youth-Series/book-series/RSCYS

Mapping the Moral Geographies of Education

Character, Citizenship and Values

Sarah Mills

Routledge
Taylor & Francis Group

LONDON AND NEW YORK

First published 2022
by Routledge
2 Park Square, Milton Park, Abingdon, Oxon OX14 4RN

and by Routledge
605 Third Avenue, New York, NY 10158

Routledge is an imprint of the Taylor & Francis Group, an informa business

© 2022 Sarah Mills

British Library Cataloguing-in-Publication Data
A catalogue record for this book is available from the British Library

Library of Congress Cataloging-in-Publication Data
A catalog record has been requested for this book

ISBN: 978-1-138-30082-8 (hbk)
ISBN: 978-1-032-10728-8 (pbk)
ISBN: 978-0-203-73306-6 (ebk)

DOI: 10.4324/9780203733066

Typeset in Times NR MT Pro
by KnowledgeWorks Global Ltd.

Contents

Acknowledgements

First and foremost, I want to thank my family and loved ones for their support with my career and always keeping my feet firmly on the ground. All of those who mean so much to me know who they are – thank you.

Second, this book represents the final stage and step beyond an Economic and Social Research Council (ESRC) Future Research Leader Award (2014–2017). I want to thank the ESRC for their financial support of my early career and especially that award [ES/L009315/I]. Although the research for this book has now taken me beyond the example of National Citizen Service (NCS), unpublished data from that project features in Chapter 7. I therefore want to gratefully acknowledge Dr Catherine Waite's role as PDRA on the linked research project that formed part of my ESRC fellowship. I also wish to acknowledge ESRC support via ES/F00737X/1, as unpublished data from that research on The Scouts now features in Chapter 5. Thanks are extended to The Scouts Heritage Service for kind permission to reference original material in Chapter 5. Furthermore, I am grateful for a School Fellowship from Loughborough University that supported research activities also underpinning this book.

Third, I am grateful to my colleagues at Geography and Environment, Loughborough University for their support and friendship. I have also been lucky enough to co-supervise a fantastic group of doctoral students during my time at Loughborough University and want to extend my thanks to Jonathan Duckett, Tim Fewtrell, Jo Hickman Dunne, Laura Crawford, Rosie Austin and Catherine Wilson.

Finally, I want to acknowledge my own geography of education and thank the countless teachers, lecturers, researchers, students and professional support staff that have shaped my journey to this point. Thank you to Milking Bank Primary School, Bishop Milner Secondary School and Sixth Form, and Aberystwyth University. A special thanks to Gornal Library in Dudley, West Midlands, where I fell in love with reading and writing.

1 Introduction

In recent years, politics in the United Kingdom has been dominated by questions of character and leadership. The 2019 General Election largely focused on individual candidates' values, moral fibre and character, despite approaches to Brexit largely leading the respective party political agendas. The character, integrity and leadership of key UK Government figures have also dominated the evaluations of its handling of the ongoing COVID-19 crisis. This salience of character in contemporary social and political life echoes a more specific trend in educational spaces over the last decade, namely, the promotion of character education.

There has been a growing trend in schools and civil society to encourage 'good character' in children and young people. This push to improve individuals, and by extension wider society and the nation-state, builds on a long historical legacy of moral education and citizenship training. Furthermore, there is a complex genealogy to ideas of character, not least in terms of religious and class-based contexts in the nineteenth and twentieth century. This book charts the recent re-imagining and re-fashioning of a 'character agenda' in England specifically, shaped by the wider political landscape of the United Kingdom. In this context, characteristics such as courage, resilience, grit, perseverance, self-discipline, responsibility, trustworthiness and neighbourliness have been woven into the pedagogic landscape, shaping moral geographies and everyday lives of children and young people.

The definition of character education is contested. However, it is largely understood as a global educational movement where learning is based on values and virtues, designed to encourage certain moral and ethical characteristics in individuals. The Office for Standards in Education, Children's Services and Skills (Ofsted) – a non-ministerial department of the UK Government – define character as: "A set of personal traits, dispositions and virtues that informs their [pupil's] motivation and guides their conduct so that they reflect wisely, learn eagerly, behave with integrity and cooperate consistently well with others" (Ofsted, 2019: 58). This book demonstrates how geography matters in understanding character education. It maps these moral geographies over time and space, tracing how character and related concepts of citizenship and values are cast in space. This book

DOI: 10.4324/9780203733066-1

traces the emergence of a 'character nation' in England over the past decade and its multi-scalar geographies. On the one hand, this national state-led push has been inspired by global examples such as the United States and several countries in Asia. On the other hand, it has been embraced through local activities in individual schools and within civil society organisations. Across a range of character education programmes and initiatives at the local, national or global scale, a number of different visions for character education exist that draw on different approaches, beliefs or pedagogical practices. Taken together, these activities reflect a growing focus on values, virtues and morals under the 'umbrella' of character, despite the nuances of philosophical and ethical traditions or scholarly distinctions. The remainder of this introductory chapter expands on the aim and objectives of the book and outlines the wider political significance of this topic, as well as provides brief chapter summaries.

Overall, there has been a shift in ideology and practice towards (or as this book argues back to) character in recent years in both formal and informal education. For example, between 2014 and 2016, the Department for Education (hereafter DfE) spent a total of £14.5 million on character education via their Character Innovation Fund and two rounds of Character Education Grants.[1] This (re)turn to ideas of moral, values or virtues-based education in England has been driven by the Conservative Government and championed by Nicky Morgan MP and later Damien Hinds MP during their respective tenures as Secretary of State for Education. Schools have been encouraged to embrace character to develop "courage, honesty, generosity, integrity, humility and a sense of justice" (DfE, 2019a: 7) through taught lessons, assemblies, partnerships with military ethos providers and extra-curricular activities. Beyond school, the language of character has been resurrected within spaces of informal education such as youth movements and outdoor educational charities. The historical legacy of citizenship training and character-building adventurous activities has been rebranded, along with a shift from youth volunteering towards 'social action'. Overall, this book explores a range of spaces and practices for children and young people designed with a 'moral compass' and situated within this wider character agenda.

From the classroom to the campsite, the book captures this shifting landscape and presents a wider argument about the relationship between formal and informal education. It does so through an analysis of historical and contemporary case studies in England including schools, military ethos providers, uniformed youth organisations and the state-led youth volunteering programme, National Citizen Service (NCS). Indeed, the originality of the research underpinning this book lies in considering spaces of formal and informal education alongside each other, considering the synergies and divergences between their ideas, practices and spatialities. This approach is vital in understanding both the historical impulses that underlie the contemporary spatialities of character education, as well as the increasingly blurred lines between schools and other learning spaces. Overall, this book traces the wider social, political and geographical implications of the character agenda in England, before examining wider global trends in schools and state-led policies on youth volunteering. This focus acts as a lens through which to view the wider relationship between character, citizenship and values.

There were three key objectives that shaped the research activities underpinning this book. First, tracing the genealogies of character education and excavating its shifting definitions, meanings and place over time. Second, critically examining the contemporary policy debate on character education in England in both formal and informal education. Finally, mapping the global geographies and geopolitics of character education, and excavating the multi-scalar politics behind its influences, formations and cultural values. In doing so, the book builds on a groundswell of recent research in human geography on childhood, youth and education, as well as long-standing work in the social sciences and humanities on educational spaces and politics. At this stage, it is important to situate the book's three central contributions to academic knowledge within the relevant wider academic debates.

First, the book's analysis contributes to debates on *geographies of youth citizenship* (Pykett, 2009; Staeheli, Attoh and Mitchell, 2013) by tracing the genealogies of character education and positioning these activities within wider historical attempts by the state and civil society to 'make' citizens (Mills, 2013; Mills and Waite, 2017) and 'govern through pedagogy' (Pykett, 2011). A body of vibrant work in children and young people's political geographies has spearheaded this specific agenda on the geographies of youth citizenship (i.e. Philo and Smith, 2003; Kallio and Häkli, 2011, 2015; Skelton, 2010, 2013; Wood, 2012; Staeheli, Attoh and Mitchell, 2013). This focus can also be situated within wider debates about young people as 'beings' or 'becomings' (Uprichard, 2008) and young people's transitions to adulthood (i.e. Valentine, 2003; Worth, 2009). The book also examines how character education has sought to create, shape and govern citizens of the future, tracing its key influences and features over time. Thus far in the literature, critical examinations of this shifting landscape within the United Kingdom remain underdeveloped, and research on character education tends to be either exclusively historical or practitioner focused and championing its benefits (i.e. Kristjánsson, 2015; Arthur et al., 2017; Arthur, 2020). Two important exceptions on character education in the United Kingdom are critical sociological work in Bull and Allen's special issue on the 'turn to character' (2018) and Jerome and Kisby's (2019) research in education studies, discussed in Chapter 2. However, this book utilises key geographical concepts, for example space, place and scale, to provide a different analytical lens to these debates in order to excavate the moral geographies of character education in England.

Second, this book contributes to vibrant and expanding work on the *geographies of education* (i.e. Hanson Thiem, 2009; Holloway et al., 2010;

Kraftl, 2013; Mills and Kraftl, 2014). Its critical examination of the geographies of character education focuses on both formal and informal learning spaces, which Holloway and Jöns (2012) observe are often studied in isolation. What is striking, and worthy of academic attention, is the increasingly blurred connections between formal and informal education via the character agenda. For example, the recent partnership between the DfE and The Scouts to fund pilot 'character' Scout Groups in schools, discussed in Chapter 3. These shifts are happening in a range of contexts, for example within alternative education provision (Kraftl, 2013) and the growing use of Forest Schools by mainstream schools in the United Kingdom (Pimlott-Wilson and Coates, 2019). An examination of such trends in relation to character, however, allows for an interrogation of the wider relationship between the state and civil society. Furthermore, by excavating the *geographies* of character education, the book highlights the unique characteristics of the United Kingdom and the impact of devolution on educational policy. In 'mapping' the synergies and distinctions between the approaches in England, Scotland, Wales and Northern Ireland, and later analysing related trends in different international contexts, this book brings an important geographical dimension to critical interdisciplinary debates on education (Brooks, Fuller and Waters, 2012; Brock, 2016).

Finally, this book contributes to work on the geographies of children, youth and families (i.e. Skelton and Valentine, 1998; Holloway and Valentine, 2000; Aitken, 2001; Hopkins, 2007; Jeffrey 2010; Holt, 2011; Holloway, 2014) by critically analysing the discourses that character education harbours about children and young people's role in society. This analysis, and how it intersects with recent moves to teach 'Fundamental British Values' for example, contributes a much-needed focus on the explicit *moral* constructions surrounding young people's 'life skills' within a body of literature that often focuses on youth employment and higher education (i.e. Holdsworth, 2017; Pimlott-Wilson, 2017). This focus also highlights the everyday geopolitics of children and young people's lives, whether through the increased presence of military ethos providers in schools or moves towards 'compulsory volunteering' under the auspices of the character agenda. Moreover, the book demonstrates how a number of these trends are extending their influence and reach towards much younger age groups, illustrating the importance of researching children's spaces. This book demonstrates the value of a geographical approach for understanding shifting yet enduring educational philosophies and agendas for children and young people, and how these are politicised and performed in everyday life.

Overall, this book has a specific focus on a set of interconnected policies and spaces of character education. More broadly though, it raises a series of provocative questions about the state's construction(s) of character, citizenship and values. Indeed, these moral geographies have wider political implications that reach far beyond school assemblies or a youth organisation's activities. For example, consider the role of 'Good Character' tests by the

Home Office of the United Kingdom as part of its citizenship infrastructures. This test was previously only required for naturalisation as a British citizen, but since 2006 it has also been applied to other routes to registration. The good character requirement utilises no set definition of character, yet recent guidance for Home Office staff states that "consideration must be given to all aspects of a person's character, including both negative factors, for example criminality, immigration law breaches and deception, and positive factors, for example contributions a person has made to society" (Home Office, 2019: 9). This formalisation of subjective assessments about the 'contributions' a good citizen makes is noteworthy. These trends also have clear impacts on the everyday lives and futures of children, young people and families. As Haque (2019) argues "hundreds of children – even those born and raised in Britain – are being denied citizenship because of minor offences". The 'Good Character' test, coupled with other moves in the hostile environment such as high fees for children's citizenship applications (Rickett, 2020), demonstrates the wider importance of mapping the moral geographies of education and this broader relationship between character, citizenship and values.

A further individualised example that demonstrates these tensions is the case of British-born Shamima Begum, whose experience reveals the fissures between character, citizenship and age-based rights and responsibilities. Begum had her British citizenship legally revoked in February 2019 following her plans to return to the United Kingdom from Syria. Begum, along with two other friends, left the United Kingdom aged 15 to join ISIL.[2] Begum has been represented in political, legal and media discourses as either an active volunteer who joined ISIL with 'adult-like' agency, or as a victim of grooming and child sexual exploitation whilst a minor. This complex legal and moral debate rumbles on and questions continue to be posed about her family, faith, background, schooling, values and character. At the time of writing, Begum, now aged 21 and based in a detention camp in Syria, is beginning an appeal process that has prompted deeper reflections on the moral boundaries of citizenship. As *Observer* columnist Kenan Malik (2020) writes, "however terrible the acts she may have committed, she remains Britain's responsibility, not the responsibility of the Kurdish forces who hold her, nor of Bangladesh, the country of her parents' birth". This example and the media spotlight is extreme, but nevertheless it highlights how constructions of childhood and adulthood are entangled with questions of character, citizenship and values. This book elucidates these wider conceptual connections and relationships through a range of different examples that map the moral geographies of education.

Overall, the book has three central arguments. First, the book argues that geography matters in understanding the character agenda, both in terms of the moral geographies that spatialise ideas about behaviour and youth citizenship, as well as the multi-scalar geographies of its practices. Second, the book argues there is a growing geopolitical dimension to character education that is vital to consider in any analysis, as children and young people's experiences of character education are increasingly shaped by military ethos and spaces. Finally, the book argues that the boundaries of formal and informal education have become increasingly blurred through the character agenda, shaping and reflecting changes in the wider relationship between the state and civil society.

The book is structured in seven remaining chapters. Chapter 2, "Moral geographies of childhood, youth and education: Learning to be citizens of good character", introduces 'moral geographies' and outlines approaches to studying childhood, youth and education within disciplinary human geography. The chapter provides an overview of the key distinctions between citizenship and character and outlines the wider philosophies that have shaped institutional spaces for children and young people. The chapter then provides a detailed synthesis of existing literature on character education across various disciplines. Finally, it outlines the data collection and analysis that underpin this book, from critical policy analysis and archival research to participatory research with young people.

Chapter 3, "Character nation: Geographies and geopolitics of education in England", focuses on the growing character agenda in England in the past decade, within the context of devolution. The chapter charts the key government policies and shifts towards a re-imagined place for character in schools and civil society, demonstrating the scalar imaginaries and global inspirations that underpinned this push. Furthermore, the chapter highlights the significant role of military ethos providers within this landscape, demonstrating a growing everyday militarism in spaces designed for children and young people. Overall, this chapter argues that the UK Government has a perennial nostalgia for educational pasts at the same time it seeks to carve out bright new educational futures.

Chapter 4 focuses on one of the core pillars of character education and its values-based understandings – grit – and its related class-based moral geographies of gumption. "Grit and Gumption" critically questions the 'new' language of character education and its promotion of stickability, resilience and preparedness by tracing these ideas over time and excavating their historical antecedents such zeal and pluck. The chapter outlines how these ideas are seen as helping children and young people to 'bounce back' from economic and social challenges driven by neoliberalism, couched within class-based constructions of confidence. It explores the role of ideal selves in formal education, mobilised through ideas of behaviour and collective school ethos that characterise the contemporary spatialities of character education. Chapter 4 also examines the extracurricular activities of schools as part of attempts to foster gritty resilience, alongside the powerful imaginative geographies of 'The Great Outdoors' utilised as natural character-building landscapes.

Chapter 5, "Be prepared, be resilient: British youth movements and civil society", provides an analysis of how moral-based citizenship training was, and remains, a key feature of uniformed youth movements in Britain. The chapter uses the detailed case study of The Scouts in the United Kingdom to demonstrate the role of civil society organisations and informal education in shaping ideas about character for children and young people. Chapter 5 specifically examines the moral geographies of camping and the Scout Farm in the early twentieth century, as part of the organisation's broader efforts to promote character-building adventurous outdoor activities. Furthermore, the chapter reflects on the contemporary push to 'be resilient' within one of the largest and most established youth organisations of the United Kingdom.

Chapter 6, "'The lessons they don't teach in class'? National Citizen Service and social action", focuses on the contemporary relationship between character, giving, generosity and neighbourliness. These ideas have recently been reframed in policy and practice as modern and new expressions of citizenly duty, exemplified by the rise of 'social action'. The chapter uses the example of NCS – a voluntary state-funded youth programme for 15–17 year olds in England and Northern Ireland – to demonstrate how moral geographies about young people's behaviours are codified as 'good' or 'bad'. Chapter 6 also outlines how NCS, as a state-led attempt to create the 'fastest growing youth movement in a century', understands both citizenship and adulthood. The chapter ends by charting the increased involvement of NCS in schools, further demonstrating the book's overall argument about the blurred boundaries of formal and informal education.

Chapter 7, "Character, citizenship and values: From national debates to global geopolitics", focuses on the wider political implications of the character agenda. The chapter critically engages with contemporary global trends on character education and identifies the growing place of youth volunteering within international election campaigns. This geopolitical analysis unpacks the oxymoron of 'compulsory volunteering' and the wider importance of understanding the relationship between character, citizenship and values. Chapter 7 reiterates the book's call that to better understand these timely and politically significant dynamics, we must consider the moral geographies that shape children and young people's lives both inside and outside of the classroom.

Chapter 8 "Conclusion" reflects on the book's central aims and arguments. This chapter demonstrates the importance of a geographical approach for understanding shifting yet persistent educational values, philosophies and activities designed for children and young people. The book ends by reflecting on the synergies between the past and present, as well as looking towards educational and political futures, not least with the unfolding challenges of COVID-19 through which ideas about character have again been expressed in significant ways.

Notes

1. £5 million via a Character Innovation Fund and £3.5 million and £6 million, respectively, on two rounds of Character Education Grants (DfE, 2017a).
2. ISIL (Islamic State of Iraq and the Levant) or in other contexts ISIS (Islamic State of Iraq and al-Sham or Syria) is a radical Islamist terrorist group.

2 Moral geographies of childhood, youth and education

Learning to be citizens of good character

Introduction

It seems somewhat ironic for a book with 'mapping' in its title that none of its pages contain any maps. This is deliberate and reflects the theoretical approach of this research, outlined in this chapter. This book's focus on moral geographies is largely situated within a body of work developed over the last 30 years as 'new cultural geography'. This approach can be traced in many ways to Jackson's (1989) book *Maps of Meaning*, where they advocated how a metaphorical mapping of signs, references and representations can reveal the contours of wider cultural politics and socio-spatial relations (on these themes and the sub-discipline's development, see Cresswell, 2010; Horton and Kraftl, 2013; Anderson, 2019). This focus on cultural politics, and key work in cultural, political and historical geography, has been a formative influence shaping the author's geographical imagination. The references to 'mapping' throughout the pages of this book are therefore in this spirit and approach.

Chapter 2 provides a conceptual overview to studying moral geographies within disciplinary human geography. More broadly, a geographical approach utilises key concepts of space, place and scale to understand the socio-spatial dynamics of everyday life. This chapter examines key literature on moral landscapes of childhood and youth, and in doing so, demonstrates how this book's specific focus on character education speaks to current academic debates. The remainder of the chapter then provides a detailed synthesis of existing research on character education from a wide range of disciplines and practitioner contexts. It ends by outlining the original data collection and analysis that underpins this book.

Moral geographies: landscapes of childhood and youth

This book sits at the confluence of, and contributes to, three specific bodies of literature introduced in Chapter 1: first, geographies of youth citizenship; second, geographies of education; and finally, geographies of children, youth and families. More broadly though, 'moral geographies' is the entry point and framework driving its analysis of character education and related debates.

DOI: 10.4324/9780203733066-2

"The constitution of what counts as moral is infused with a geographical imagination and shot through with ideology" (Cresswell, 2005: 128; see also Cresswell, 1996). Research has sought to excavate these dynamics through examining the moral geographies of policies, practices and in/exclusions across society, space and time (i.e. Driver, 1988; Matless, 1994), in tandem with broader questions of morals, morality and ethics in geography (i.e. Smith, 2000; Barnett, 2012). As Legg and Brown observe in their excellent review, "the majority of work claiming the title of 'moral geography' focuses on codes of conduct and the regulation of human behaviour through spatial relations" (2013: 135). This focus has characterised a vein of geographical research on cities, science, leisure, land use, institutions and spaces of childhood and education (explored fully below). Mapping such geographies of moral regulation, which are often shaped by axis of social difference and power, connects to wider ideas of citizenship, belonging, landscape and nationhood. For example, David Matless' analysis of the 'art of right living' (1995) demonstrated how landscape can be entwined with a moral sense of citizenship and used as a method and motif to 'improve' the citizen. His research on leisure in the English countryside during the 1930s and 1940s illustrated how "always...the good citizen made sense only in relation to a contrasting 'anti-citizen'" (1997: 142; see also Matless, 1998). This 'other' of countryside visitors, including urban dwellers in motor cars and walkers, was needed by planners and preservationists not only to display both a lack of citizenship, but also to generate new senses of citizenship in the public (see also Merriman, 2005; Parker, 2006 on the Country Code). This focus in human geography on how moral codes are expressed and 'take shape' is important to recognise here, before this section's specific focus on the moral geographies of childhood, youth and education.

A relatively small body of work in cultural-historical geography has been instrumental in revealing the place of moral regulation in educational spaces. This has most notably been realised through archival research on schools in different national contexts that demonstrate their role (and indeed geography's own role) in making and shaping young citizens (i.e. Ploszajska, 1994, 1996, 1998; Gruffudd, 1996; Cameron, 2006; de Leeuw, 2009). Beyond school, other spaces designed by adults for children are also infused with moral geographies (Aitken, 2001). This is demonstrated by Gagen's (2000) instructive research on playgrounds in the United States in the early twentieth century shaped by ideologies of gender and national identity (see also Gagen, 2004a). Several scholars have demonstrated how educational spaces are influenced by wider social constructions of childhood, for example the powerful construction of children as either 'angels or devils' (Valentine, 1996) or wider idea(l)s and moral landscapes of childhood and youth (Gagen, 2004b; Kraftl, 2006). In turn, spaces such as schools are then shaped by children's own identities and their everyday lived experiences and geographies (Holloway and Valentine, 2000).

The connections between spaces of childhood and learning have been central to other work in this field, notably where codes of conduct have spatialised behaviour and educational training has been linked to wider notions of good citizenship and national values. For example, research that has explored the 'place' of moral geographies in the emergence and growth of uniformed youth movements in Britain during the late nineteenth and early twentieth century, inspired by different imperial, religious or political projects. The author's previous research on the Scout Movement (Mills, 2013), Girlguiding (Mills, 2011a), Jewish Lads' Brigade (Mills, 2015) and Woodcraft Folk (Mills, 2016) demonstrates how youth movements have been intrinsically linked to moral projects as spaces of youth citizenship. These organisations have sought to train children and young people to become the good (future) citizen for the nation and wider global citizenry. At the same time, there was a clear emphasis in their educational programmes on behaviours in the present, whether in uniform or performed at home. As spaces with a moral compass, behaviour was observed and regulated, for example at the campsite or the meeting place, as part of these wider institutional geographies. Crucially though, these spaces were shaped by children and young people themselves. These moral geographies of youth organisations have always raised wider questions about the *character* of their membership – and by extension – the character of the youthful population of the nation at that time. The spatial politics of uniformed youth movements are elucidated further in Chapter 5. However, the temporalities of youthful citizenship underpinning these dynamics, which frame the overall book and its focus on character education, are outlined in detail in the next section.

Being/becoming citizens and being/becoming adults: the place of education

Definitions of citizenship are contested but generally refer to an individual's membership of a political community (Staeheli, 2011; Yarwood, 2013). Citizenship formations – especially for children and young people – have changed over time and are shaped by geography, connected to wider ideas of identity and belonging (Isin and Wood, 1999). The position and framing of children and young people as "citizens-in-the-making" (Marshall, 1950: 25) reflects many approaches and ideas about the place of children and young people and their relationship to the state, invoking ideas of rights and responsibilities, participation and futurity.

It has been well acknowledged that childhood and adulthood are social constructions that are spatially and temporally specific (Jenks, 1996; Holloway and Valentine, 2000), with children and young people held in a constant tension between 'being' and 'becoming' in the transition to adulthood (Valentine, 2003; Uprichard, 2008). However, it is important to focus on the institutional geographies (Philo and Parr, 2000) that formalise these 'moments' of transition towards a productive adulthood, embodied in the figure of the 'good' citizen. What attributes are seen to be constitutive of 'good citizenship'? How is this taught? Indeed, as well as a time of socialisation into 'adulthood', young people are also seen as needing to learn how to be *citizens* in the future. It is this process of 'learning to be a citizen' which frames and drives many of the questions interrogated in this book.

Ruth Lister et al. (2003)'s dichotomy between the law-abiding and trouble-making citizen is useful here in understanding citizenship formations in the context of children and young people. Lister et al. (2003)'s interviews with young people living in the United Kingdom revealed that many understandings of the nature of citizenship were connected to set ideas about good and bad behaviour. The idea of "law-abiding and non-disruptive versus the trouble-making citizen" (2003: 245) is used to frame and essentialise young people, highlighting how models of behaviour are used as frameworks for understanding the norms of good citizenship (see also Pykett, Saward and Schaefer, 2010). More recently, scholars have further conceptualised these ideas of behaviour and belonging, notably in relation to debates on migration (Anderson, 2013) and stigma (Tyler, 2020), revealing their enduring power. Indeed, the politics and performance of citizenship formations for children and young people are often intertwined with these wider debates.

The notion of 'learning to be a citizen' (Staeheli, 2018) and education's role in this context is shaped by (adult) questions concerning what kind of citizen young people ought to become, or conversely, what type of (uncivil) citizen they would become if not guided in the 'correct' way. This concern about future citizens and the 'youth debate' has, as Bill Osgerby argues, come "to function as an important ideological vehicle that encapsulates more general hopes and fears about the state of the nation" (1998: 1). Take, for example, author Douglas Hubery's fears over the future moral character of Britain in the 1960s:

> The question, pointedly stated in the King George's Jubilee Trust Report [1955], *Citizens of Tomorrow,* is not 'How much do they know?' or 'What more can we teach them?' It is 'What sort of human beings are they going to be?
>
> (1963: 59)

These types of polemic questions can still be heard in contemporary social and political life. On the one hand, fears are expressed over the decreasing civic, political or democratic participation of young people (Putnam, 1995; Buckingham, 2000) or their 'troublesome' or declining immoral behaviour (Griffin, 2001; Roche et al., 2004). On the other hand, youth is mobilised as a powerful metaphor for hope in the face of anxieties, insecurities and unknown possible futures (Katz, 2008; Kraftl, 2010). These hopes and fears are in constant flux, spatialised through institutional geographies, and most powerfully through education.

The drivers for education are diverse but underlying most formal and informal learning spaces are attempts to foster hope and allay such fears, instilling good citizenship for the nation's future. Geographers have demonstrated these ideas across a range of global examples, such as the role of education in the United States as part of a quest to enhance its economic competitiveness (Mitchell, 2018), as a wider nation-building and security project in South Africa (Staeheli and Hammett, 2010) or as an arena for reworking civic inclusion for indigenous youth in Chile (Webb and Radcliffe, 2015). The geographies of education are therefore vital to consider in debates on youth citizenship and are a vibrant body of scholarship more broadly (i.e. see Collins and Coleman, 2008; Hanson Thiem, 2009; Holloway et al., 2010; Holloway and Jöns, 2012; Mills and Kraftl, 2016). This book's examination of the recent politicisation of character education in England can be situated within this wider scholarship.

In the United Kingdom specifically, the notion of *active* citizenship for children and young people is important to outline at this point, particularly in relation to debates about being/becoming citizens and being/becoming adults. Active citizenship is the active participation of individuals acting upon their rights and responsibilities, rather than citizenship as a purely legal framework (Kearns 1992, 1995). The UK Government has been involved in fostering or cultivating active citizenship for children and young people in two central ways. First, via a 'curricular' pathway of formal school lessons and teaching materials. For example, citizenship education was introduced to the National Curriculum in English schools in 2002 (QCA, 1998; Heater, 2001; discussed in Chapter 3). Second, active citizenship has been encouraged by the state via 'co- or extra-curricular' activities by schools either in partnership with, or run independently through, youth charities and organisations. For example, a range of voluntary youth programmes in civil society are premised on the ideology of active citizenship, and the state has increasingly sought to 'co-opt' them into their formal educational remit. Active citizenship has been a key idea in this space across the United Kingdom for the last 30 years, for example in participatory spaces such as school councils and other forums focused on young people's voices (Matthews, 2001; Percy-Smith, 2010) and a wide range of service learning and volunteering programmes in the United Kingdom (Mills and Waite, 2017). These spaces have encouraged young people to participate in various activities – primarily at the local scale – in order to become active responsible citizen-subjects; an emasculated version of citizenship while a young person waits for the 'real thing'. The final part of this section therefore outlines the role of both formal and informal education in this context and key scholarship in these areas, before moving to consider character education specifically.

There has been some excellent scholarship on citizenship education in schools in England (i.e. Weller, 2007; Pykett, 2009, 2011; Keating et al., 2010; Mycock and Tonge, 2011a; Weinberg and Flinders, 2018) and on shifts towards governing emotional literacy in formal education (Gagen, 2015). These moves are precursors to the character agenda examined in this book, and there is of course a global landscape of citizenship education (i.e. Kerr, 1999) beyond this specific national focus. It is also important to contextualise character education in relation to recent educational restructuring in England within a wider context of neoliberalism, inequality and social reproduction (Holloway and Pimlott-Wilson, 2011; Reay, 2017; Mitchell, 2018; Holloway and Kirby, 2020), as well as pushes for democratic or political education (Sloam, 2008; Suissa, 2015) and teaching 'Fundamental British Values' (Starkey, 2008). A detailed discussion on each of these debates in relation to formal education appears in relevant sections of Chapters 3 and 4.

Beyond school, informal education also plays a key role in 'learning to be a citizen'. The philosophies and practices of informal or non-formal education (Jeffs and Smith, 2005) are entangled with wider histories and geographies (Mills and Kraftl, 2014). Geographers have interrogated how spaces such as youth work (Blazek, 2015; Dickens and Lonie, 2016), alternative education (Kraftl, 2013), exchange programmes (Fairless Nicholson, 2020) and extracurricular enrichment activities (Holloway and Pimlott-Wilson, 2014a) are important sites of friendship and socialisation for children and young people, yet shaped by wider geographies of class, race and gender. Furthermore, academic research from youth studies, criminology, and by youth workers themselves, has outlined the increased politicisation of youth work through state-funded initiatives. For example, through the targeting of youth work to certain groups of young people and ideas of participation, citizenship, activism and impact (Williamson, 1993; Wood, 2010; Davies, 2018; de St Croix, 2018).

What is striking about the character education agenda, and drive's much of this book's argumentation, is the increasingly blurred connections between formal and informal education. As later chapters in this book demonstrate, both curricular and co- or extracurricular activities are used to promote character education, including:

> assemblies, subject lessons, dedicated character education lessons, sports, performance arts clubs, outward bound activities, hobby clubs and subject learning clubs. These opportunities help young people to explore and express their character and build the skills they need for resilience, empathy and employability
>
> (DfE, 2019a: 6; see also DfE 2017a)

The connections and relationships between formal and informal educational spaces – often studied in isolation – require further attention (Holloway and Jöns, 2012). Important exceptions on these blurred boundaries include Pimlott-Wilson and Coates (2019) on forest schools, Cartwright (2012) and Davies (2014) on youth work in schools and Kraftl (2013) on alternative education. Nevertheless, this book importantly examines specific sites of character education within both formal and informal learning spaces, providing a still relatively unique cultural-historical approach within the locus of work on geographies of youth citizenship, geographies of education and geographies of children, youth and families.

From citizenship to character? Situating the debate

This section specifically considers character education and how key definitions and understandings of this concept differ from citizenship education. It reviews extant scholarly work on character education from education studies, psychology, history and sociology, as well as practitioner contexts. Indeed, despite the vibrancy of existing work, character education has not specifically been focused upon by geographers to date. Furthermore, as debates about young people in the United Kingdom and their political participation are now arguably moving away from citizenship education towards character education (Weinberg and Flinders, 2018), it is important to focus on these emerging and influential policies and spaces.

As Chapter 1 hinted, a standard definition of character education has proved elusive, with diverse expressions and understandings of this movement across the world. It is essentially a focus on values or virtues-based learning that is designed to encourage, teach or 'unlock' certain moral or ethical characteristics. However, for others, the term character education has been used more generally to capture efforts that promote 'soft' or 'life' skills and wider personal development. This complexity (or vagueness) and lack of a singular definition have been acknowledged by the UK Government in their own policy documents and infrastructures, which will be discussed in Chapter 3. In terms of the etymology of character itself though, this is often defined as an 'essence' or used to describe certain traits, attributes and characteristics. It captures where one's actions align with one's values and where ethical and moral ideas are enrolled with certain virtues. The cultural historian Warren Susman (1984) argued that the nineteenth century was the 'age of character', before the age of personality which defined the twentieth century (see also Allen, 1986 on this shift). There is often a relationship drawn between questions of national character and questions of individual character (Mandler, 2007), but it is clear that ideas about this 'essence' – at an individual subject level and more broadly – have shifted over time and space.

There are divergent, and at times controversial, views on whether the 'essence' of character education is based on *virtues*, as per the Jubilee Centre for Character and Virtues at the University of Birmingham, UK, discussed shortly, or *values* as per organisations such as the values-based Education Trust. Character education as moral *virtues* focuses on the positive moral attributes associated with character, for example 'gratitude' (Carr, 2016) and good habits for a 'successful' life. As Harrison and Bawden state "virtues, as we understand them, are settled (stable and consistent) traits of character, concerned with morally praiseworthy conduct in specific (significant and distinguishable) spheres of human life" (2016: 16). This view of character education through a virtue ethical perspective proposes that such virtues can be learnt (Kristjánsson, 2015; Arthur, 2020). It is this theoretical understanding that has, to date, dominated much of the renewed character education focus in England, attributed by Jerome and Kisby (2019) to an active policy community that emphasises individualised private or personal ethics rather than public ethics. It is important to recognise that an alternative approach considers character education through the lens of *values* (on these debates and taxonomies of character education see Lickona, 1992; Kohn 1997; Howard, Berkowitz and Schaeffer, 2004). This foregrounds more diffuse and contested ideas such as respect, responsibility, kindness, fairness, honesty, compassion and courage. In both a virtues or values approach to character education, which do overlap at times in practice, these ideas are often linked to moral norms and faith-based contexts, for example its relationship to religious studies in schools (Metcalfe and Moulin-Stozek, 2020). The role of religious and faith-based communities in schools and youth work is long-standing and significant (Feinberg, 2006; Hemming, 2015; Mills, 2015). For example, the recent debates in the United Kingdom on faith schools (Valins, 2002; Dwyer and Parutis, 2012) as part of wider educational restructuring in the United Kingdom (discussed in Chapter 3) bring these geographies and politics into stark relief. More broadly, the role of Christianity in shaping moral education in colonial and postcolonial contexts across the world is vital to recognise. This has been the focus of much important scholarship, tracing how such political and religious dynamics have had significant legacies in educational practices and politics, particularly in sub-Saharan Africa (White, 1996; Brock-Utne, 2000) and Latin America (Klaiber, 2009). In addition, religion has often defined the politics of character education in the United States (Howard, Berkowitz and Schaeffer, 2004) and is drawn upon at relevant parts of this book's discussion.

Overall, whether based on virtues or values-based learning, character education is more than merely encouraging certain personality traits or skills. It has deeper moral geographies that shape its philosophies and pedagogies, which this book excavates in the remaining chapters. This book is not concerned with advocating for a particular 'take' on character education to be adopted and does not propose a preference or specific theorisation of character education in philosophical or ethical terms. Indeed, it often uses values and virtues interchangeably to reflect the approach, discourses and materials of the UK Government and many youth organisations. The book's central message and argumentation is that the spatial expressions of these moral ideas have wider implications and that geography matters. For example, the slippage in ideas about behaviour and morality used in criminal and youth justice contexts, or the Home Office's 'Good Character

Test' introduced in Chapter 1. Indeed, these expressions and infrastructures of values-based citizenship have fuelled controversial and at times violent manifestations of character-based judgements and punishments.

Academic research on character education spans several disciplines and practitioner contexts. In education studies, developmental psychology and political philosophy, character education is often discussed interchangeably with moral education (i.e. Nucci and Narvaez, 2008; Althof and Berkowitz, 2006). Some similarities and differences are still outlined, but nevertheless the *Journal of Moral Education* (1971–present), published via the Association for Moral Education, and the *Journal of Character Education* (2003–present), both publish research that explores character, moral, ethical, spiritual, cultural and civic education, with articles covering a wide range of geographical contexts from the 1800s onwards. In this book, the term 'character education' is used, given the UK Government and Department for Education's (DfE's) own exclusive use of this term. Nevertheless, it is important to highlight the academic debate surrounding these definitional dilemmas as part of a wider interdisciplinary project.

A more clear-cut distinction is the one between character and *citizenship* education. For Osler, "character education focuses on the moral and the individual ... [whereas] citizenship education focuses on the collective, and has both moral and political dimensions" (2016: 13; see also Osler and Starkey, 2005; Ajegbo, 2007). Jerome and Kisby (2019) fully outline the differences between both concepts within the United Kingdom, strongly advocating for citizenship education rather than a character education which they argue is "best viewed as deeply flawed in both theory and practice" (2019: 3). They also highlight how champions of character education "tend to treat the good person and the good citizen as essentially synonymous" (2019: 20) and they propose there should be a clearer distinction between these two ideas. Davies, Gorard and McGuinn (2005) also caution about the perceived similarities between the two and that educators need to be cognisant of their contrasting meanings. However, others such as Carr (2006) have stressed that the relationship between citizenship and character is more complex and nuanced, for example when one considers the moral roots of citizenship. This book is not a direct comparison between character or citizenship education, but rather it maps the moral geographies underlying character education and examines its spatial expressions in England. In doing so, this analysis reveals the wider contours of character, citizenship and values, as well as the wider (geo)political significance of these ideas within and beyond education.

In the United Kingdom, there is a long and complex history to character education. The idea that children and young people need a wide-ranging education that nurtures certain aspects of 'good', positive and virtuous elements of character is not new (Taylor, 2018; Arthur, 2020). Important scholarship by historians of education has excavated how character-building activities have been woven into educational spaces since the nineteenth century. For example, English social reformer William Ellis (1800–1881) promoted a focus on character in the establishment of several schools (Stewart, 1972), and these pedagogical histories and individual biographies are important to recognise. Character became a very popular idea in early twentieth century Britain, entwined with masculine ideals. For example, in his book *The Making of Character* first published in 1900, John MacCunn described character as "the manly and man-making duties of local and imperial citizenship" (1931 [1900]: 124), and these gendered dynamics are explored later in the book. Historians have demonstrated how throughout the twentieth century, the classroom and other spaces in civil society have encouraged moral education, influenced by educators and other organisations. For example, Wright's (2017) research on the role of secular activists who sought to shape the education of working-class children in English schools between 1897 and 1944. This work is instructive in outlining the key role of organisations such as The Moral Instruction League and Association for Education in Citizenship who shaped practices within educational spaces. The 1944 Education Act in England is often used as a milestone in the history of education and itself referenced the nation's 'most abiding assets' as 'the character and competence of a great people' (cited in Barber 1994: 77; see also McCulloch, 2011), as well as highlighting the role of outdoor education (Cook, 1999; see also Edwards, 2002 on geography's own disciplinary role in 'values education'). These ideas were typified by the growth of organisations such as the Outward Bound Trust (Freeman, 2011), an educational charity that works closely with schools. Such moves were given new life during this period, fuelled by earlier beliefs across Europe in the role of leisure, sports and pursuits for community-building and participation in 'civic life' (Baker, 2017).

Overall, the popularity of character education in the United Kingdom has waxed and waned between the nineteenth and the twenty-first century, yet it does currently find itself 'back in fashion' (Arthur, 2005). The recent revival of character education by various champions in UK Government and influential networks, outlined in Chapter 3, has led to a flurry of recent research in education studies and sociology, sparking a range of responses. On the one hand, some researchers have critiqued the contemporary political philosophies of character education and its co-option by various interest groups and individuals (i.e. Suissa, 2015) building on longer standing critiques of the movement (i.e. Kohn, 1997). On the other hand, some researchers have advocated for the expansion of character education, attempting to quell fears, scepticism and quash myths that it is old-fashioned or paternalistic (i.e. Kristjánsson, 2013). The key research centre working on character education in the United Kingdom – which has now become an object of research in itself – is the Jubilee Centre for Character and Virtues at the University of Birmingham, UK. Opened in 2012, its research and activities are multi-disciplinary and practitioner-focused, engaging teachers and providing distance learning courses. For the Centre, character education dates back to Aristotle and has not only a complex historical trajectory (Kristjánsson, 2015; Arthur, 2020) but also a strong contemporary value that should not be dismissed (Harrison and Bawden, 2016).

As hinted, the Jubilee Centre itself has been critically analysed for the role it plays within wider policy and philanthropic networks (Allen and Bull, 2018; Bull and Allen, 2018). Furthermore, its theorisations of character education as well as its educational resources and practices have been subject to detailed critical analysis (Jerome and Kisby, 2019). Nevertheless, the Centre is comprised of a wide body of scholars with diverse views in this field and has conducted some of the most extensive research on contemporary character education in the United Kingdom to date. For example, its 2015 report *Character Education in UK Schools* engaged with over 10,000 students and 255 teachers (see also Arthur et al., 2017). Its publications do tend to largely be targeted at a professional and practitioner audience, calling *for* character education and supporting UK schools with relevant resources. In contrast, research on character education in sociology and critical education studies has tended to highlight its role within wider debates on governance (Spohrer and Bailey, 2020), with important ethnographic research demonstrating how teachers in the United Kingdom reproduce, negotiate and in some cases refuse its dominant discourses and activities (Morrin, 2018). Overall, these critical engagements and academic debates hint at a topic riddled with complexities, which are returned to at various parts of this book when relevant.

The final element to this review of existing literature is to highlight scholarship on the development of character education in different international contexts. This research has largely focused on the United States where character education is arguably viewed as a science across a wide range of organisations and networks (i.e. Berkowitz, 2002; Berkowitz and Bier, 2005). Character education's more recent growth in the United States has been fuelled by the charter school movement as part of widespread public service reform and social and political insecurities (Winton, 2008; Saltman, 2014; Gawlik, 2016). Its recent popularity is part of a much wider historical trajectory of general support for character education (Howard, Berkowitz and Schaeffer, 2004) and moral education in the United States (McClellan, 1999), rooted in a focus on educational development and virtues (Edmonson, Tatman and Slate, 2009). Indeed, the influential American reformer John Dewey (1859–1952) promoted the moral mission' of schools and his wider educational and social reform often hinged on a moral imagination (Dewey, 1944; Fesmire, 2003; Pietig, 2006). However, Winton (2008) traces the contested history of character education in the United States during the twentieth century and its patchy evidence-base and studies. More recently, contemporary spaces such as KIPP schools (Knowledge is Power Program), a network inspired by positive psychology, are indicative of the continued and re-envisioned growth of character education in the United States (Angrist et al., 2012). Smagorinsky and Taxel (2005) argue that this debate reflects broader 'culture wars' in the United States, and Love (2019a) has powerfully outlined character education's place in a wider racist structure of what they term the 'educational survival complex' in the United States.

A number of countries in Asia also have strong historical and contemporary relationships to character education and have been the subject of detailed academic studies. For example, Roesgaard's (2017) examination of moral education in Japan outlines both the historical place of values and their place within recent changes to the school curriculum and teaching materials. Chapter 7 of this book focuses on a range of contemporary global examples beyond the United States and Asia as part of an overall discussion on the geographies and geopolitics of character education. Overall, the book has not only a clear analytical focus on England, situated within the landscape of the United Kingdom, but also locates its discussion and arguments with reference to multi-scalar geographies and international connections.

Methods and data analysis

The final section of this chapter outlines the data collection and analysis underpinning this book. During an Economic and Social Research Council (ESRC) Future Research Leader fellowship award (ES/L009315/I) that hosted a research project on National Citizen Service (NCS) in the United Kingdom (introduced shortly), the wider character agenda was gaining pace and permeating spaces of formal and informal education. This sparked further research activities on historical and contemporary insights into the geographies of character education, including critical policy analysis and archival fieldwork, as well as the mixed methods data set outlined below. The specific case study examples drawn upon in this book are introduced in the relevant respective chapters, but it is important to highlight here that the book draws on three sets of research material.

First, critical policy analysis was conducted to examine UK Government policies on character education and related debates in the last decade, as well as interrogate their wider constructions of children and young people. This fieldwork focused on key documents from 2010 to 2020 including DfE press releases, reports and speeches about character education. Extensive related government policies over the last ten years were analysed such as the *Giving White Paper* (2011) and other relevant documentation produced by the UK Government, think-tanks, consultancies and the third sector. This data analysis primarily informs the arguments presented in Chapters 3 and 4 on the character agenda in schools. These written sources were analysed via critical discourse analysis to explore how "interconnected groups of texts, statements and representations" about children, young people and education emerge and 'take place' (Kraftl, Horton and Tucker, 2012: 15). This approach allows for the moral geographies of character education to be mapped in the spirit of the approach outlined earlier in this chapter.

Second, the book draws on historical data collected and analysed via archival fieldwork. This includes unpublished archival data originally collected during research projects on uniformed youth organisations. These research projects focused primarily on citizenship, gender, religion and nation-building within the respective case studies (i.e. Mills 2013, 2015, 2016). However, sources and material on *character* were also analysed for a potential future research project and now find their home in this book, especially in Chapter 5 on The Scouts. This rich material has been coupled with further archival fieldwork at the National Archives, London, UK, to examine wider state discourses on youth citizenship which overlapped and combined with notions of character at key flashpoints. Through locating contemporary policy directions within such wider histories, the book elucidates how character has been defined and understood by the state and in civil society, and how this moral impulse has been cast in space.

Finally, the book draws on unpublished data from the aforementioned ESRC research project on NCS – a UK government-funded youth programme launched in 2011. The recent focus on character within this voluntary programme for 15- to 17-year olds in England and Northern Ireland is noteworthy and presented in Chapter 6. The original quantitative and qualitative fieldwork includes 8 semi-structured interviews with key 'architects' of NCS, 23 semi-structured interviews with Regional Delivery Providers who delivered the programme in England and Northern Ireland (2011–15) and an online survey with 407 NCS graduates who participated in the programme between 2011 and 2015. A sample of these NCS alumni (30) was later interviewed providing further qualitative data. The final methodological component was a four-week ethnography of one NCS cohort in one region in Summer 2015 and a linked participatory animated whiteboard-video project. This was written and directed by young people from the ethnographic research but produced by a professional animation company. In addition to the acknowledgments at the start of this book, sincere thanks are expressed to Dr Catherine Waite for her fieldwork contributions as a Postdoctoral Research Associate (PDRA) within the NCS project. The wider ESRC fellowship also supported the author's ongoing analysis of the emerging geopolitics of character education and 'compulsory volunteering', discussed in Chapter 7.

Taken together, this new research material on the renewed and revived character agenda and the strands of fieldwork activity outlined above inform the book's overall arguments. The book does not draw on quantitative or qualitative data with teachers or children in schools, or make claims about these lived experiences; indeed, the previous section highlighted original existing scholarship in this area. Young people's own voices and lived experiences are still very much present in this book through the chapter on NCS. Overall, this book combines those insights alongside careful and critically engaged policy and archival research to unpack the geographies and geopolitics of character education in schools and civil society.

Conclusion

This chapter has introduced the study of moral geographies and situated scholarship on landscapes of childhood, youth and education within disciplinary human geography. It defined character education and provided some historical context to ideas around 'learning to be a citizen' and spaces constructed as arenas for moral instruction. The chapter has also reviewed a growing body of academic literature on character education in disparate disciplines, before outlining the data collection and analysis underpinning this book.

This book does not seek to propose its own philosophical or ethical 'take' on character education. Indeed, it is not a study based within educational psychology or evidence-based evaluations of character education programmes. Rather, it maps the moral geographies of a re(turn) to character in England and uses this example as a lens through which to view a series of interconnected debates on the wider relationship between character, citizenship and values. Indeed, the book demonstrates the 'place' of character in contemporary social and political life. It is also worth stating that the book neither argues for or against character education in schools or beyond. This may disappoint some readers, however the polemic extremes of this debate have tended to characterise the literature to date and character education has supporters and critics across the whole political spectrum. Rather, this book focuses on what the presence, practices and politics of character education tell us about the geographies of education and youth citizenship more broadly. Specifically, it crafts three key arguments about the geographies of character education, the geopolitics of character education and the increasingly blurred boundaries between formal and informal education.

A final reflection in this chapter is to stress that despite the clear value in working with both historical and contemporary research material to understand spaces of childhood and youth, challenges remain. Mapping the moral geographies of character education through the diverse source material outlined in the previous section requires careful and reflexive practice, whether handling material from interviews with young people or historical documents from over a century ago. In the remaining chapters, the rest of the book seeks to retain a keen historical sensitivity to broader notions of character, citizenship and values, locating contemporary policy directions within these wider histories. It turns now to the multi-scalar geographies of character education hinted at thus far, framed through the vision of a 'character nation'.

3 Character nation

Geographies and geopolitics of education in England

Introduction

This chapter examines the philosophies and spatialities of character education in England. Despite the research on character education outlined in Chapter 2, there is a need for a geographical approach that captures the sites, spaces and multi-scalar expressions of its recent revival. Indeed, scale is an important conceptual device in this book used to understand and explain the moral geographies of character education. Scale can be used as a metaphor and conceptual device to explain how ideas and practices occur at (and between) the local, national and global scale (Herod, 2009). Geographers have advocated the value of scale in the study of education (Purcell, 2011), youth citizenship (Wood, 2012; Mills and Waite, 2017) and moral geographies more broadly (Legg and Brown, 2013). As Chapters 1 and 2 have hinted, the aim of this first analytical chapter is to excavate the moral geographies of character education's recent re(turn) in England and to demonstrate how geography matters through a multi-scalar analysis. In doing so, the chapter contributes to the central arguments of the book on the geographies and geopolitics of character education, as well as the blurred boundaries of formal and informal education.

The chapter begins by charting the emergence of a renewed character agenda in recent years, critically analysing contemporary policies and frameworks for character education. These reveal wider constructions of childhood and youth, operating at different spatial scales. This chapter goes on to place the rise (or return) of character education in England in the wider context of educational reforms in the United Kingdom and a landscape of devolution. The chapter ends by critically examining the place of military ethos providers as part of attempts to foster a 'character nation'.

Champions of character and cabinet shuffles

The Conservative Party have introduced several educational policies and reforms in the last decade. They returned to power as part of a coalition government with the Liberal Democrats (2010–2015) led by David Cameron

DOI: 10.4324/9780203733066-3

as Prime Minister. This period was followed by successive Conservative Party Governments led by Theresa May PM with a party majority between 2015 and 2017, and then in partnership with the DUP between 2017 and 2019. Since the 2019 General Election, and at the time of writing, Prime Minister Boris Johnson leads a Conservative Party majority. Michael Gove MP was David Cameron's first Secretary of State for Education (2010–2014) and led a largely ideological drive for stronger curriculum knowledge as well as expanding the role of academy schools, discussed later. His successor, Nicky Morgan MP (2014–2016), mostly continued Gove's policies but also became a central champion for character education.

This focus on character education was cemented in the £5 million Character Innovation Fund announced as a 'landmark step' by Morgan in December 2014. This was launched to support the development of character in schools and focused on grants, awards and evidence-based research, including funding eight projects that would "use military ethos in schools to improve education attainment of the most disengaged pupils" (DfE, 2014a). A key focus in Morgan's 2014 announcement were the DfE Character Awards where schools could apply to be recognised as leaders in character education if they could demonstrate "their programme develops character traits, attributes and behaviours that underpin success in school and work" (DfE, 2015a). These traits included "perseverance, resilience and grit; confidence and optimism; motivation, drive and ambition; neighbourliness and community spirit; tolerance and respect; honesty, integrity and dignity; conscientiousness, curiosity and focus" (DfE, 2014b). Even in this initial list, one can identify the emergence of virtues and values formalised as part of this Department for Education (DfE)-funded programme. This book's analysis primarily focuses on these original discourses and claims, specifically unpacking the moral geographies of 'perseverance, resilience and grit' (Chapters 4 and 5) and 'neighbourliness and community spirit' (Chapter 6). However, over time, the number of perceived attributes that character education fosters has grown. Indeed, Jerome and Kisby (2019) identified 68 different characteristics within DfE-written and-funded publications that character education is seen to promote. The central attributes cited in the 2014 launch therefore form part of a much wider, diverse and at times ambiguous set of related characteristics.

Two key drivers for character education in the run-up to this launch were the 'Character and Resilience Manifesto' published by the All-Party Parliamentary Group (APPG) on Social Mobility (Paterson, Tyler and Lexmond, 2014) and a wide-ranging Riots Communities and Victims Panel following riots in a number of English cities in August 2011 (Briggs, 2012). The APPG on Social Mobility's manifesto has been examined in detail by Burman (2018) and Spohrer and Bailey (2020), but it is important to note here as part of the genealogy of character education in the United Kingdom. Second, the Riots Communities and Victims Panel advocated a stronger role for character education a few years earlier, stating that "the key to avoiding future riots is to have communities that work...where parents and schools ensure children develop the values, skills and character to make the right choices at crucial moments" (Riots Communities and Victims Panel, 2012: 6). The report referred to Ofsted's role, "Personal, Social, Health and Economic education (PSHE, discussed later in this chapter) and the need for 'regular assessments of pupils' strength of character" (2012: 8). The report's constructions of young people were that those involved lacked moral fibre rather than a recognition of structural inequalities or a desire to protest police violence. These two reports, whilst not the central driver for character education that came from Morgan's stewardship, are still important to acknowledge.

The emergence of Morgan's Character Awards in England was largely focused on recognising existing good practice in schools. The awards of £15,000 for 27 schools (with a £20,000 prize for the national winner) sought to highlight those schools whose approaches included:

> integration [of character] into the curriculum and wider aspects of a school; the teaching of character as a separate subject; extra-curricular activities, such as sport and music; and outward facing activities, such as community work and volunteering.
>
> (DfE, 2014b)

Building on the 2014 Character Innovation Fund, £3.5 million was later announced by the DfE for Character Education Grants, cementing Morgan's vision and intended future direction. These 14 projects for the 2015/6 academic year included new approaches to character education as well as evaluations of existing schemes, with awards made to organisations including St John Ambulance (£254,911), PSHE Association (£137,000), Premiership Rugby (£556,494), The Prince's Trust (£584, 366) and the Youth Sport Trust (£95,527) (DfE, 2015b; see also Birdwell, Scott and Reynolds, 2015). One of these DfE pilots, as an indicative example, was a new collaboration with The Scouts (£302,299) called 'character by doing' (Scott, Reynolds and Cadywould, 2016). This example is worthy of more discussion for its reshaping of the geographies of education, blurring the boundaries between formal and informal learning.

The pilot project introduced scouting in six schools in Leicester, Boston, Oxford, Thurrock, Swanley and Brighton with a programme linked to the Key Stage 2 (KS2) national curriculum. An evaluation of the scheme found that "students participating in the pilot improved their leadership capabilities by 22 per cent over six months" as well as improved behaviour and better school attendance, yet there were also barriers and challenges (Demos, 2016). Although a relatively small-scale pilot, this example is significant because it marks, for the first time, the UK Government directly funding The Scouts, a voluntary uniformed youth movement discussed in full in Chapter 5. Furthermore, the nature of this pilot programme was that scouting activities were not always delivered by volunteers on week-day evenings in an extracurricular format, but sometimes via teachers or teaching assistants within the school day. This *blending* of formal and informal education is noteworthy. Here, an organisation that has always been voluntary was translated into a DfE-funded programme under the umbrella of 'character'. My interest here is less with the pilot's outcomes, impact or evaluation (see Jerome and Kisby, 2019: 103–4) but rather what this funded pilot's format tells us about the wider re-shaping of education in Britain. This example demonstrates one of this book's core arguments about a more fundamental shift that has taken place in the relationship between formal and informal education, and by extension, between the state and civil society.

Following the 2015 General Election, Nicky Morgan continued to pursue character education and more explicitly linked this to other aspects of Conservative Party policy, such as social mobility, stating:

> Character Education is part of our core mission to deliver real social justice by giving all children, regardless of background, the chance to fulfil their potential and achieve their high aspirations
>
> (DfE, 2015b)

This vision was expanded the following year, with a further £6 million of DfE funding for projects and programmes in another round of Character Education Grants. Importantly, one-third of these projects funded in 2016 had a 'military ethos approach' to character, explored later in this chapter.

Morgan's push for character education was cut short however following the appointment of Justine Greening MP as Secretary of State for Education in 2016. This followed Theresa May's appointment as Prime Minister after David Cameron's resignation in the aftermath of the EU referendum. Morgan went on to write a book on character education (2017), which largely focuses on examples of best practice in schools. She continues to champion for its expansion, including briefly as Secretary of State for the Department of Digital, Culture, Media and Sport in Boris Johnson's Cabinet (July 2019) before moving to the House of Lords.

Greening scrapped the character education pilots and awards, diverting the funding to the UK Government's 'Opportunity Areas Scheme'. In a move away from Morgan's 'pet project' of character (Whittaker, 2019a), Greening focused on 12 deprived opportunity areas, with Schools Minister Nick Gibb announcing in 2017 that the character education programme had closed. This was replaced by a £22 million Essential Life Skills programme targeting disadvantaged children, yet still included the promotion of "extra-curricular activities, such as sports, volunteering and social action projects" (cited in George, 2017). Greening's tenure as Education Secretary between 2016 and 2018 instead focused on new areas of educational policy such as a national funding formula for schools and primary school assessment.

However, after Greening's resignation during another cabinet reshuffle in January 2018, the appointment of Damian Hinds MP as Secretary of State for Education saw a revival of character education. In his first big speech in this role, which he delivered as part of the Church of England Foundation for Educational Leadership Conference, Hinds focused on character and declared that education "is about more than academic achievement" (DfE, 2019b). The following year, he cemented this focus by formally launching the DfE's 'Five Foundations for Building Character': Sport, Creativity, Performing, World of Work, and Volunteering & Membership.[1] These aspects are discussed more fully later in the book, but it is important to note that responses to this announcement from many teachers, certainly on social media networks, highlighted existing work in this area that had been overlooked, and funding challenges in these exact areas. Teaching organisations and unions stressed the wider structural issues of teacher retention and exam pressures that hamper these five elements flourishing. Indeed, running alongside all of the above developments on character education has been an education system, and families, who have been impacted by a climate and reality of austerity (Ridge, 2013; O'Hara, 2014; Hall, 2019).

To return to Hinds' key speech, here he announced the re-launch of the DfE Character Awards and an audit of out-of-school activities and volunteering opportunities, as well as a new Character Advisory Group. This group of experts chaired by Ian Bauckman led a DfE consultation on 'character and resilience' in Summer 2019, during which time Hinds was replaced by Boris Johnson, the new incoming Prime Minister. Gavin Williamson MP was appointed as Secretary of State for Education and remains in this post at the time of writing. Although Williamson has not publicly 'backed' character education with the same verve as Hinds or Morgan, an official DfE character education framework was published in November 2019 and reflects the same direction of travel.

The state's vision for character education has increasingly been formalised within policy apparatus and its wider approach to governance. Hinds was clear in February 2019 that the DfE's focus on character education:

> is not about a DfE plan for building character. It has to be about schools learning from other schools, it's about business pitching in when it can, it's about community groups speaking up and inviting schools in. It's about individual adults volunteering. All of us need to work together using the wide range of resources and experts that there are out there.
>
> (DfE, 2019b)

This vision from Government reflects wider Conservative Party agendas for a 'Big' and then 'Shared' Society (Mills and Waite, 2018), whereby the state emphasises the responsibilities of individual citizens yet where infrastructures of welfare are simultaneously rolled back (see Mohan, 2012; Painter, 2012; see also Chapter 6). However, despite this rhetoric away from the DfE management of character, there have been recent moves towards benchmarks for its measurement. These are less stringent than Gatsby benchmarks used by schools to rate their own careers advice (Whittaker and Murray, 2019) with no requirement to submit data on character education explicitly (Whittaker, 2019b). However, the most recent Ofsted Framework introduced in September 2019 states that "inspectors will evaluate the extent to which schools support pupils to develop their character – including their resilience, confidence and independence" as part of a focus on "personal development and positive attitudes" (DfE, 2019a).

This section has outlined how support for character education has fluctuated in the last decade during a very turbulent time in British politics. However, it is worth stressing that it has not just been the Conservative Party who has championed character education. In 2014, the same year as Morgan's launch of formal character education infrastructures, Shadow Education Secretary Tristram Hunt MP of the Labour Party argued that resilience and character should be taught in schools, strongly linked to creativity (Demos, 2014). The idea has some notable cross-party appeal. For example, a 2015 poll of 150 MPs across all parties revealed 77% agreed "developing a sense of moral values is as important for school children as good GCSE and A-level results" (The Jubilee Centre, 2015). In more recent years, the Labour Party has focused its opposition policies and manifesto pledges on education around a 'National Education Service' and 'cradle-to-grave' free learning. It is therefore the Conservative Government that has primarily driven the character agenda whilst in power, and the next sections of this chapter elucidate this multi-scalar geographical imagination.

National visions and global ambitions

The place of character education within the DfE in recent years can be understood through a scalar-based framework. This section presents two key observations that demonstrate the role of national visions and global ambitions, drawing from a critical analysis of Morgan and Hinds' speeches and other relevant published material and documentary sources.

First, within the state's vision, Britain (read England) was cast as a potential 'global leader' of character education. Morgan outlined at the launch of her multi-million-pound character education programme that:

> It will cement our position as a global leader in teaching character and resilience, and will send a clear signal that our young people are being better prepared than ever before to lead tomorrow's Britain
>
> (DfE, 2014c)

The aim was clearly to create a character nation that prepared young citizens for 'life in modern Britain' (DfE, 2014c), yet the global scale was invoked through ideas of leadership and educational prowess. In suggesting that Britain would become the 'best' at teaching character, despite the long-standing expertise and experience of other countries outlined in Chapter 2, one can read the kind of jingoistic bravado and colonial nostalgia that a number of researchers have identified as a feature of the recent UK Government, especially in relation to Brexit and a 'Global Britain' (Beaumont, 2017; Bhambra, 2017; Daddow, 2019; Dorling and Tomlinson, 2019). The lofty and overly ambitious aim to be a global leader in character education via a small number of pilot projects chimes with other characteristics of the Conservative Party in recent years, for example attempts by Michael Gove during his time as Education Secretary to rise in the global educational rankings. This desire to catch international competitors such as China and Singapore in a race of global educational performance fuelled his reforms to secure higher-achieving academic attainment by UK pupils (You and Morris, 2016). This educational 'battle' between the perceived 'softer' work of character education, and the 'harder' work of academic achievement (see Breslin, 2016), has therefore been framed as areas where the UK Government has sought to become a global leader.

Second, an analysis of Morgan and Hinds' speeches and related DfE material reveals how character education was justified and framed by UK Government in relation to national economic stability and growth. Character education would have a direct economic benefit to the nation, as well as providing young citizens with individual moral fibre. As the DfE stated in 2014:

> Creating a strong work ethic, along with raising educational standards and improving classroom behaviour, is also crucial for building a strong future for Britain's economy
>
> (DfE, 2014c)

The transitions from school to employment are highlighted numerous times across relevant material, revealing a wider economic rationale for character education as well as ideological ideas about nation-building and espousing global ambitions. Application submissions from schools to the Character Awards scheme had to "evidence that their activities result in better grades, improved behaviour or improved job prospects" (DfE, 2014d). These instrumentalist ideas about 'making workers' (Mitchell, 2018) and competing in a global economic marketplace are notable. Indeed, Nicky Morgan returned to this rationale in her later book, stressing that "employers are putting an ever greater premium on character traits such as resilience, persistence, grit, leadership, self-awareness and self-efficacy", with a bold claim that "the English education system is only doing half the job it needs to do to prepare our children for the 21st century" (2017: 4). Clearly, preparedness for the labour market to foster national economic growth has been an important factor in this multi-scalar vision for character education.

The discussion in this section so far has excavated the geographical imaginations that frame England's recent moves towards character education. It is also worth stressing at this stage how a number of international examples were positioned as global inspirations in this national quest. For example, the DfE highlighted the work of the King Solomon Academy in North West London, where the headteacher:

> has introduced much of the thinking of the American charter school movement and specifically that of its Knowledge is Power program (KIPP), where schools have longer days and a structured approach to developing and talking about character.
>
> (DfE, 2014a)

This movement in the United States, introduced in Chapter 2, is often cited as inspiring some English schools, who are then showcased by DfE. As well as global influences, international examples of character education and partnerships are encouraged by DfE. These are mobilised through a number of global networks and interest groups, explored in Chapter 7. These are more strategic partnership style relationships, rather than activities couched in a sense of 'global citizenship'. Indeed, literature on citizenship education has highlighted the scalar tensions in this area. For example, Divala and Enslin (2008) outline in their study on Malawi how most assumptions and practices of citizenship education still focus on the nation-state rather than discourses of global citizenship education (see also Nieto, 2018 on these debates in Latin America). However, there are other examples where civic education is used by many countries to stretch the boundaries of the nation-state towards ideas of global citizenship, empowerment and rights-based discourses, such as in Argentina (Suárez, 2008). Over the last decade, the move in England towards (or back to) character has instead been framed as an ideological and economic national vision, couched within global ambitions and drawing on selective international inspirations.

This section has illustrated the scalar-based geographical imaginations used to frame this recent push for a character nation. The next element to this chapter's multi-scalar discussion is to demonstrate how the practices of character education are expressed and performed at the local scale, specifically through three 'pathways' in schools. The next section foregrounds the local and everyday geographies of character within formal education, placing character education in the context of wider and related educational policies in England and the United Kingdom.

Local and everyday geographies: three pathways to character

Character education is widely understood as either explicit or implicit within schools in England. The recent focus on virtues and values has been expressed and performed through three central pathways of activity at the local scale. First, as 'taught' within the classroom; second, via a whole-school ethos or values-based approach to school life; and finally, 'caught' through extracurricular activities.

These three pathways, examined in detail in the remainder of this chapter and Chapter 4, are evident in how the DfE understands character education. It frames character education as a collective endeavour with no set fixed root. For example, in one of their first announcements on character education in 2014, they state:

> Character can be developed in pupils in a wide variety of ways, through teaching values in personal, social, health and economic (PSHE) or citizenship lessons, through the full curriculum, or by competing on the playing field or taking part in extra-curricular activities such as the Duke of Edinburgh Award, the National Citizen Service or after school debating clubs.
>
> (DfE, 2014a)

This quote links the development of character within individual pupils at the local scale to existing national infrastructures within the curriculum, discussed shortly, as well as national organisations in civil society such as the Duke of Edinburgh Award founded in 1956 and the more recent UK Government programme National Citizen Service (NCS), discussed in Chapter 6. This quote captures the three pathways, but also reflects Breslin's observation that policy makers have dealt with the personal and social development curriculum in an "everywhere but nowhere" hinterland of cross-curricular themes (2016: 21).

It is therefore important at this point in the chapter to position character education in relation to a series of wider educational infrastructures and policies. Education in England is generally structured into primary schools (usually for children aged 5–11[2]) and secondary schools (usually for those aged 11–16), with geography and other socio-economic factors shaping school 'choice' (Butler and Hamnett, 2010). This landscape is comprised of free publicly funded state schools, historically referred to as comprehensives, as well as fee-paying independent schools, often referred to as private or elite schools. A number of those retain a historic name and status through the oxymoron of a 'public' school (Gamsu, 2018). Furthermore, there are also state-funded selective schools, or 'grammar' schools (Gorard and Siddiqui, 2018), where entrance is based on academic performance tests. Finally, there are also a growing number of academies and free schools in England. Academy schools were introduced in England in 2000 under Tony Blair's Labour Government, allowing forms of privatisation and choice within public education through external sponsorship by charities, business and faith communities. This academisation of schools has been analysed as an entrepreneurial programme (Woods, Woods and Gunter, 2007) representative of the wider neoliberal restructuring of education (Purcell, 2011).

Its relevance to this section's discussion is that some of the recent flexibility at the local scale to embrace character education (or not) has been due to the continued expansion of academy schools and the introduction of free schools as new privately run but state-funded schools (Hatcher, 2011). This is particularly striking in relation to the second pathway of 'school ethos', whereby some schools have completed reframed themselves around the 'buzzword' of character. This wider and increasingly autonomous educational system therefore hosts a range of diverse approaches to character education at the local scale, with even Morgan stating that "the specific virtues to be developed are best left to each school and its community" (2017: 36).

Another key factor shaping the 'place' of character at the local and everyday level within individual schools is its non-statutory subject status. The National Curriculum, an overarching framework for schools established in 1988, does not currently include character education. Indeed, it is promoted by the DfE as a route to achieving many other statutory and non-statutory elements of formal education including Citizenship Education, 'Personal, Social, Health and Economic education' (PSHE), 'Fundamental British Values' (FBV) and the most long-standing focus in this area, 'Spiritual, Moral, Social and Cultural Education' (SMSC). These four areas, outlined in turn shortly, mean that the first pathway of character education – taught within the classroom – has generally involved not so much the teaching of character as a separate subject (although see Arthur et al., 2017), but rather its integration within other subject lessons dependent on local and individual school choices.

Citizenship education in England has been enshrined in the National Curriculum since 2002 and remains a statutory subject in secondary schools and non-statutory framework in primary schools (Davies, Gorard and McGuinn, 2005; Weller, 2007; Pykett, 2009, 2011; Keating et al., 2010; Mycock and Tonge, 2011a). Its distinction from character education was introduced in Chapter 2, and this related field was recently described by the House of Lords Select Committee on Citizenship and Civic Engagement as in a 'parlour state' (House of Lords, 2018; see also Weinberg and Flinders, 2018). The latest guidance for PSHE, a more long-standing feature of formal education in the United Kingdom since the 1990s, describes it as a non-statutory subject, but "an important and necessary part of all pupils' education" (DfE, 2019c). The PSHE Association describe how the subject develops "skills and attributes [that] help pupils to stay healthy, safe and prepare them for life and work in modern Britain" (PSHE Association, 2020). More recently, PSHE has been under the spotlight given amendments in September 2020 to include compulsory 'health education and relationships education' in primary schools and 'relationships and sex education' (RSE) in secondary schools, creating its own political and ethical debates on 'virtues' (see Bull, 2019a). The most controversial of the four educational policies discussed here as curriculum-based 'routes' to achieving character is the teaching of FBV. FBV are defined as 'democracy, rule of law, individual liberty and mutual respect and tolerance' (DfE, 2014d: 5) and have been a compulsory part of English schools since their inclusion in Teachers' Standards in 2012. This move has been critically examined by education scholars and sociologists for links to the wider counter terrorism 'Prevent' duty (DfE, 2015c; Lander, 2016; Smith, 2016; Starkey, 2018), seen as part of a wider ideological quest for 'Britishness' with clear racial and religious overtones. Finally, there is a longer-standing focus in the English education system on the "social, moral, spiritual and cultural development" (SMSC) of pupils "to prepare them for opportunities, responsibilities and experiences of later life" (DfE, 2014e: 5). The duty of SMSC has been enshrined in English education since 1988 (Peterson et al., 2014) with many newer initiatives often framed as sitting within this SMSC 'umbrella'. For example, the 2019 Ofsted Education Inspection Framework includes a focus on FBV and character with explicit reference to SMSC. Overall, the educational policies outlined here (and returned to in the next section on devolution) espouse different yet interlinked cultural values for children and young people and are one part of the wider apparatus through which individual schools at the local scale can focus their character education activities.

A key finding of the critical policy analysis underpinning this chapter is how between 2014 and 2020, individual schools at the local scale have been 'held up' as exemplars of character across the three central pathways. For example, Wellington College, an independent school in Berkshire, has had timetabled lessons in resilience since 2006 and is often cited as a key example of best practice in the classroom (BBC, 2014). Second, King's Leadership Academy Hawthornes, Merseyside is included as a case study in the 2019 DfE Character Education Framework for its use of a military ethos model, exemplifying the second pathway of whole-school ethos. Finally, the third pathway around extracurricular activities shines through in examples such as Bonus Pastor Catholic College in Greater London who introduced "The Charter" of 60 enrichment experiences and extracurricular opportunities which "helps to develop [pupils] determination and resilience" (DfE, 2019a: 23). Overall, examples of the local and everyday geographies of character education 'on the ground' have been utilised by DfE, and particularly through Hinds' speeches in more recent years, to demonstrate how successful character initiatives operate and function through individual school, teacher or pupil efforts. Indeed, Morgan described such schools as "not waiting to be told to prioritise character education" (2017: 15).

The push for a 'character nation' outlined earlier in this chapter is ultimately institutionalised through the three central pathways identified in this section and individual school choices. These can include, but are not limited to, assemblies, in-class activities, reward systems, guest speakers, sports, mentoring, or co- or extracurricular opportunities. A number of schools in England take this further through embodying character education more forcefully via visual signs and markers in the school environment such as awards or mission statements. These often manifest themselves as slogans, murals, poems or acrostics of virtues and values. For example, the acronym BCL 'Building Character for Learning' in Babington College, East Midlands, or DIRT in King's Langley, Hertfordshire, which stands for 'Dedicated Improvement and Reflection Time' (cited in Morgan, 2017: 56–7). These mottos and motifs extend to younger age groups, with MAGIC at Archibald Primary School in Middlesbrough, which stands for "Motivation, Attitude, Gumption, 'I Learn' and Communication" (cited in Walker, Sims and Kettlewell, 2017: 9). These examples of the second pathway around whole school ethos, explored in depth in Chapter 4, are also supported and sustained by national networks. For example, schools in England can choose to apply to the Association of Character Education, a non-profit membership organisation, for kitemarked 'School of Character' status as part of their wider work "to develop and promote character education responses that enable young people and societies to flourish" (ACE, 2020). As such, some individual Headteachers and schools have 'hooked' onto the recent character agenda and become champions at the local level, just as Morgan and Hinds were champions at the Cabinet level. The local scale is therefore crucial to the actual lived expression of the formal policy documents and wider agenda outlined thus far. However, individual teachers do negotiate these ideas, with non-compliance at a local level too. Morrin (2018) importantly demonstrates these dynamics through her ethnographic research at a secondary school in Northern England, where character is couched in an 'entrepreneurial' spirit. They illustrate how "on one hand, the character initiative is embedded and complied with, but, on the other hand, teachers' practice is also littered with instances of 'refusal' and non-compliance" (2018: 459). Overall, the three pathways at the everyday local scale are therefore variable in the extent to which teachers and students engage with them, revealing complex geographies of education.

Devolution

An important and often overlooked element of character education's geographies is the impact of devolution. Education in the United Kingdom is shaped by devolved governance following the establishment of a National Assembly for Wales, Scottish Parliament and Northern Ireland Assembly in the late 1990s (Goodwin, Jones and Jones, 2005). The educational landscape across the United Kingdom has also been shaped by more established traditions and histories in England, Wales, Scotland and Northern Ireland (Phillips, 2003). This has led to some key differences in educational structures, age-ranges and programmes, most notably in Scotland (Bryce and Humes, 2003), and it is important to be cognisant of these dynamics ahead of the book's remaining discussion which primarily focuses on England.

Character education in the United Kingdom has primarily been an English endeavour and has not been 'taken up' with gusto either philosophically or in practice in any of the devolved nations. For example, the Character

Awards or pilots infrastructure outlined earlier has not been re-envisioned in Wales, Scotland or Northern Ireland. Conversely, citizenship education has subsequently developed with different and divergent expressions in each of the 'home nations' (Andrews and Mycock, 2007). The remainder of this section outlines two wider points on devolution, emphasising the multi-scalar geographies at play in the construction and maintenance of a renewed character agenda.

First, it is important to stress that although the revived focus on character education outlined in the book thus far has largely played out in England alone, there are 'character-esque' educational policies operating within Wales, Scotland and Northern Ireland. For example, the Curriculum for Excellence (CfE) in Scotland implemented in 2010 which is "intended to help children and young people gain the knowledge, skills and attributes needed for life in the 21st century" (Scottish Government, 2019). The full curriculum makes no reference to character, however it does refer to its purpose in helping children and young people become "successful learners, confident individuals, responsible citizens, and effective contributors" (Scottish Government, 2019). In addition, the Modern Studies curriculum introduced in Scotland in the early 1960s already includes political civic education (Andrews and Mycock, 2007). Northern Ireland's own curriculum in schools, first introduced in 1992, has been further developed since devolution and has had various iterations, but without direct explicit reference to character education to date. However, there has been an important focus on the role of citizenship education in this context (Smith, 2003), currently manifest in a unit on 'local and global citizenship'; indeed, this debate in education studies is often discussed in tandem with developments in the Republic of Ireland, which hosted Civic, Social and Political Education (CSPE) between 1997 and 2019 (Nugent, 2006). Finally, the new Curriculum for Wales will be launched in 2022, and its draft at the time of writing does not explicitly use the word character. It does, however, describe how the learning experience should support "characteristics, attributes and values of wider skills" including 'be resilient' and 'be persistent' (Welsh Government, 2019), which chime with ideas of character education in England, expanded on in Chapter 4. Overall, the developments in Scotland, Northern Ireland and Wales are related but distinct, hence the book's core focus on England for its pursuit and embrace of character education but with a keen awareness of these devolved geographies.

Second, the geographies of devolution in the context of education are not limited to schools but also powerfully shape NCS. This voluntary government-funded youth programme was launched in England and later expanded to Northern Ireland. However, it does not run in Scotland and despite being piloted in Wales in 2014, was not pursued by Welsh Government. As Chapter 6 outlines in full, NCS is increasingly seen as a pathway to develop character. However, it is important to highlight here that it has also been politicised in the context of devolution. For example, in August 2019 the then Secretary of State for Northern Ireland Nick Hurd MP stated at an NCS ceremony in Belfast that young people should "make their voices heard to politicians and get Stormont up and running".[3] This reference to a recent political stalemate in Northern Ireland demonstrates popular constructions of young people as representing hopeful political futures, especially in devolved politics.

Overall, there is a grand *global* vision used to frame the UK Government's *national* drive for character, couched in ideas about shaping modern Britain and its future citizenry. However, in reality, these lofty aims have only been invested or taken up within England due to the *devolved* politics of education, and with a patchy and divergent uptake by *local* schools who choose the extent to which they embrace this agenda via optional pathways. This fractured uptake to date is likely to change with the cemented focus on character in the new Ofsted framework and growing moves to measure and assess schools on their engagement. In the final substantive section of this chapter, the analytical focus turns from the multi-scalar geographies of character education to an explicit consideration of its geopolitics.

Military ethos providers: from geographies to geopolitics

One of the most significant and controversial aspects of the recent character agenda in England has been its relationship to military philosophies and infrastructures. The everyday geopolitics and place of military ethos in the drive to forge a character nation reflects what Basham (2016) has identified as a wider 'elevation' of military values in austere Britain. Furthermore, this example demonstrates how the multi-scalar geographies examined in this chapter thus far also include the scale of the body through embodied (masculinist) performances of characterful, duty-bound service.

Military ethos projects were described in one of Nicky Morgan's first character education announcements as having "a positive impact in improving behaviour, attendance and resilience" (DfE, 2014a). This statement can be traced back to the findings of a small-scale DfE-funded review of military ethos programmes for 'alternative provision' (Clay and Thomas, 2014), released the day before Morgan's character education announcement. Alternative provision (AP) is defined as "any education provision and/or supporting activities outside the mainstream school system...arranged by schools or local authorities for pupils who, because of exclusion, illness or other reasons, would not otherwise receive education provision" (Clay and Thomas, 2014: 10). Basham has astutely shown in her study of military ethos initiatives that specifically target working-class boys how the AP programme harbours ideas of 'self-reliance' within a Conservative government vision "to render bodies politically useful" (2016: 265). The 2013–2014 review of military ethos provision, which also included pupils referred to

AP to improve their behaviour, resilience or confidence, found there were positive outcomes for 'personal character' including self-respect and respect for others (Clay and Thomas, 2014: 7). The research focused on six AP providers who were awarded a total of £8.2 million by the DfE between 2012/3 and 2013/4. These included not only well-known UK charities such as the 'Prince's Trust' but also newer organisations such as 'Challenger Troop', which draw more heavily on a military ethos. Although hard to define, the DfE and related policy documentation described military ethos as a focus on "teamwork, self-discipline, self-confidence and leadership" (Clay and Thomas, 2014: 11).

The significance of the above review is that a number of military ethos programmes involved in the AP funding programme were later announced as recipients of the first round of DfE-funded Character Education Pilots, expanding their work beyond AP contexts. This included Challenger Troop, who were awarded £1 million for their leadership engagement programmes designed for "vulnerable or disengaged pupils aged 8 to 16 across the UK, particularly in the toughest areas of London and the South East" (DfE, 2014a). The use of the word 'toughest' by the DfE here is notable, marking a shift change towards highly masculinised language and away from terms such as 'socially deprived' or 'hard-to-reach' young people. Another scheme – 'Commando Joe's' – was also awarded £1 million for its "military-style educational teambuilding" (DfE, 2014b: 18) with "trained instructors and challenging school-focused activities" (DfE, 2014a). Mike Hamilton, Director of Commando Joe's, stated that "We're focusing on fostering altruism, bounce back, comfort zone busting and determination in school children, all of which ultimately boost attendance and attainment" (DfE, 2014a). These ideas of gritty resilience embodied as 'bounce back' are fully explored in Chapter 4. It is worth highlighting here though that these same terms are part of an 'A–D' of military ethos, introduced as the 'building blocks of character' by the DfE in their original visions for this wider work in 2014:

A: altruism – including helping others through volunteering, understanding how their behaviour affects others, helping out and home and in the classroom

B: bounce back – learning from your mistakes, developing grit and determination, overcoming failure and trying again

C: comfort zone busting – trying out new activities in unfamiliar environments, collaborating with pupils from other schools, working with new people

D: destination – establishing high aspirations and doing well at school, setting goals and understanding how to get there, developing links with employers, achieving qualifications and skills beyond the classroom

(DfE, 2014a)

These concepts are framed in policies and speeches as helping all pupils – not just those in alternative provision – "do well in school and beyond" (DfE, 2014a). However, as Jerome and Kisby (2019) highlight, there is a lack of rigorous evidence on the impacts of such schemes, despite their dominant success in the early character education funding rounds. Nevertheless, the grants awarded to military ethos providers outlined here demonstrate their place at the heart of the 'character nation' vision. This element, as we saw in the previous section, is also communicated through local examples of best practice within school walls. For example, Morgan (2017) recounts the role of 'Sergeant Redrop', PE teacher and mentor at Gordana School in South West England (2017: 44). He was employed via the US-inspired 'Troops to Teachers' scheme (Basham, 2016) and had employed various character-building interventions with disengaged pupils. Indeed, the use of global inspirations again works in and through these policy narratives and spaces. For example, the 2012 Riots Communities and Victims Panel discussed earlier in this chapter specifically refers to the US-Army programme 'Master Resilience Trainer' and its focus on 'emotional fitness' and character as well as physical education (2012: 51).

In the United Kingdom, school connections to military ethos providers were also rewarded in the 2015 Character Education Awards. For example, the King's Leadership Academy in Warrington, England, was one of 21 schools awarded £15,000, plus an additional £20,000 prize, for various character-based initiatives that notably included encouraging its pupils to join the Army Cadets. Nationally, the Army Cadet Force is aimed at 12- to 18-year-olds and has two strands: a community-based Army Cadet Force and school-based Combined Cadet Force. The recruitment of Army Cadets, especially through online materials, is heavily shaped by ideas of masculinity and social class (Wells, 2014). There is a historical legacy of Army Cadet Units in some schools in the United Kingdom, but this recent encouragement via character education is unprecedented and has been controversial (Whittaker, 2016). During his time as Secretary of State for Education, Damien Hinds detailed a 'Cadet Expansion Programme' to increase the number of Cadet Units in schools to 500 (DfE, 2019b). Although technically the British Army state that the Cadet infrastructure is "not part of the recruiting process for the Armed Forces", they continue that "we do however promote an understanding of what the Armed Forces' roles and responsibilities are, and provide assistance to any cadet who expresses an interest in joining the Armed Forces later in life" (British Army, 2020).

This section has outlined the growing militarisation of educational spaces for children and young people in England within the character agenda. First, through the role of military ethos providers within AP and subsequently the character education pilots, shaping the experiences of thousands of children and young people. Second, the more widespread encouragement for schools to establish Army Cadet Units. The philosophies and practices of these two strands of activity, centred on the embodied (masculinist) performances of individual disciplined bodies, are noteworthy. It suggests a growing need to be cognisant of the critical geopolitics shaping children and young people's everyday lives (Benwell and Hopkins, 2016) and the often 'ludic' or 'playful' dimensions to these spaces and activities (Woodyer and Carter, 2020). The latest appointment of Gavin Williamson MP as the current Secretary of State for Education, as an Army Veteran and the former Defence Secretary, suggests that a focus on military ethos looks set to continue as part of the state's wider vision to create a character nation.

More broadly, the relationship between military ethos and character education outlined in this section demonstrates the role of nostalgia in educational policies and spaces. This chapter's analysis suggests a clamoured attempt to 'return' British youth (and by extension the nation) to a set of values or characteristics that are apparently lost. This nostalgic desire to recapture an imagined sense of discipline or moral fortitude is incredibly powerful, fuelling the contemporary articulations and spatialities of character education. These emotive cultural politics have been the driving force for a series of attempts to educate children and young people in the last century, whether through uniformed youth movements (Chapter 5) or the fashioning of a modern 'national service' scheme for teenagers (Chapter 6). There is a constant tension between the past and present in educational spaces, fuelled by long-standing moral panics. For example, Theresa May's short-lived and ultimately failed attempt for a 'new generation' of grammar schools (Asthana and Campbell, 2017). This sought a partial return to the name, ethos and 'character' of this educational system of the 1950s and 1960s, phased out in the 1960s and 1970s and then banned in 1998 under Tony Blair. These recent examples therefore reveal how there is a wider desire to return to, or recapture, certain idea(l)s, morals, values and virtues within current and future generations. Indeed, the UK Government has a perennial nostalgia for educational pasts at the same time it is trying to carve out bright new educational futures. The recent pursuit of character education by the UK Government outlined in this chapter has largely been framed as a 'new' idea. However, character education has a long and complex genealogy which is excavated further in the forthcoming chapters.

Conclusion

This chapter has critically examined the multi-scalar geographies of character education in England in recent years. This chapter has traced how an agenda for character, articulated through national visions and global ambitions, has ebbed and flowed in popularity during a tumultuous period in British politics. The infrastructures, frameworks and practices that have followed are ultimately shaped by the devolved and local geographies of education, and by individual schools in England. Their leadership's decisions whether to embrace or bypass character education and its three central pathways (lessons, school ethos or extracurricular activities) create a complex landscape, further interrogated in the remainder of this book.

The multi-scalar analysis presented in this chapter supports the book's first central argument that geography matters in understanding the character agenda. Although the focus here has largely been on its multi-scalar geographies, the explicit moral geographies of character education are the focus of Chapter 4 and more forcefully build on this chapter's discussion. Next, the focus on military ethos providers in this chapter has illustrated the early contours of this book's second argument that the pursuit of a 'character nation' in England has an important geopolitical dimension. This example, and other pilots discussed earlier in this chapter such as the partnership between DfE and The Scouts, also supports the book's third argument that the boundaries of formal and informal education have become increasingly blurred. The examples of a range of organisations and charities operating within this policy and funding landscape prompt critical questions about the wider relationship between the state and civil society, and the extent to which activities are still voluntary when such actors have an increased presence in schools and formal education.

To return to Nicky Morgan, a central figure in this chapter, she stated that "we don't want to set down rigid guidelines on [character] because character isn't a one-size-fits-all concept. It isn't just one thing. It's a combination of traits that set people apart so they can achieve their dreams" (2016, cited in Morgan, 2017: 13). The inference to social mobility in this quote, and the suggestion of better futures as the direct result of possessing a stronger character, lies at the heart of a debate discussed in Chapter 4. This next chapter critically examines how grit and gumption have emerged as key ideas in this context of achieving and developing 'better character'. Indeed, Chapter 4 excavates the moral geographies of resilience and bounce-back as part of the wider quest to create (or restore) a 'character nation'.

Notes

1. Department for Education (2019) Twitter Post @educationgovuk, 7 February 2019. Available from https://twitter.com/educationgovuk/status/1093474116047421440.
2. There are also early years foundation Key Stages in England, usually for children aged 3–4.
3. Northern Ireland Office (2019) Twitter Post @NIOgov, 21 August 2019. Available from: https://twitter.com/NIOgov/status/1164259793634254848?s=03.

4 Grit and gumption

Introduction

This chapter examines grit and gumption as key ideas in the recent drive to create a character nation, framed as a way of helping young people deal with life's 'ups and downs'. This chapter charts related discourses of resilience, perseverance and preparedness that are mobilised to provide children and young people with the ability to seemingly 'bounce back' from economic and social challenges or personal hardships. These ideas are increasingly captured through the term grit, whose place in the character education agenda is worthy of specific attention.

The recent institutionalisation of grit can be partly traced to Angela Duckworth's successful popular psychology book *GRIT*, published in 2016. Secretary of State for Education Damien Hinds referenced Duckworth's book in his landmark speech of February 2019, discussed in Chapter 3, recommitting the Department for Education (DfE) to character education after its hiatus under Justine Greening. In her book, Duckworth advocates perseverance, the "daily discipline of trying to do things better than we did yesterday" (2016: 91) as a life philosophy. However, coupled with this admirable quest is advice to "strive to be the grittiest" (2016: xv) with examples across the book from the world of business and sport about hard work and not making excuses. Duckworth's book, TED talk and ideas on grit have become incredibly popular with supporters of character education and networks. However, Arthur, Kristjánsson and Thoma (2016) posed the question of whether grit really is 'the magic elixir of good character', concluding that the jury was still out. Indeed, Credé (2018) identifies a range of mixed responses to grit from educational researchers and practitioners. There is no doubt though that grit is a central idea within the character education landscape, for example its place within recent DfE guidance, speeches and other material introduced in Chapter 3. Its role contributes to Jerome and Kisby's (2019) wider assessment that character education in Britain at present has an individualistic focus. This chapter does not evaluate grit as a specific method of educational psychology, but rather it critically examines how grit has become embedded in the

DOI: 10.4324/9780203733066-4

character education agenda in England, often through class-based moral geographies of gumption. In doing so, the chapter continues to build the book's argumentation on the geographies and geopolitics of character education and the increasingly blurred boundaries between formal and informal learning spaces.

This chapter begins by mapping the terrain of debates on grit, social mobility and confidence. It then outlines some powerful historical ideas and features of school life that have been resurrected in the current landscape of character education in England such as reward systems and school ethos to shape the formation of gritty citizen-subjects. This chapter then outlines how schools in England, particularly through their co- or extracurricular activities, are increasingly utilising outdoor natural landscapes to encourage bounce-back and perseverance. Overall, this chapter demonstrates the contours of moral geographies centred on grit and gumption that underlie and fuel the wider character agenda.

'Bounce-back' and social mobility

Grit, gumption, resilience and perseverance are some of the most popular words in the wider data set underpinning this book. Resilience had been promoted by DfE for a number of years prior to the 2014 announcement of the Character Innovation Fund, but the increase in references to grit in key recent material, and the inferences about social mobility, is striking and the focus of this section's discussion.

A gritty and dogged determination to succeed is seen as achievable through the three pathways of character education outlined in the previous chapter: first, 'taught' within the classroom; second, via school ethos; and finally, 'caught' through extracurricular activities. References to grit initially emerged in the DfE's focus on military ethos, discussed in Chapter 3, and specifically as part of "B – bounce back –– learning from your mistakes, developing grit and determination, overcoming failure and trying again" (DfE, 2014a). The term then gained traction via the popularity of Duckworth's book, introduced earlier. The 'bouncebackability' of young people, to draw on a phrase often attributed to football manager Iain Dowie, reflects a wider feature of contemporary neoliberal life. The following definition of neoliberalism is helpful of "a worldview (or philosophy of life), a wide-ranging policy programme, and a set of concrete policy measures in which individual freedom is given priority" but neoliberalism is "a complex historical-geographical formation marked by unevenness and variety" (Castree, Kitchen, and Rogers, 2013: 339–340). Gill and Orgad (2018) have identified how within a neoliberal context, resilience as a 'regulatory ideal' features heavily in mainstream popular culture. Their analysis of predominantly European and US-based women's magazines, self-help books and smartphone apps reveal the "classed and gendered dimensions of injunctions to resilience" (Gill and Orgad, 2018: 477). Indeed, some have described

Duckworth's book on grit as "written in the style of self-help books" (Arthur, Kristjánsson and Thoma, 2016). The significance of self-help discourses in the lives of children and young people has also been highlighted by Cranston (2017) through her examination of its role in shaping the identities of 'Third Culture Kids' (TCK). They argue that the TCK industry, focused on children and young people that grow up 'globally mobile', can be read in terms of the production and consumption of self-help, for example through a specific TCK book, which can then shape and govern emotions and subjectivities.

In terms of character education, ideas of resilient bounce-back have increasingly filtered into educational spaces in the United Kingdom through initiatives such as the 'Art of Brilliance' course, which describes bounce-back as a 'modern day superpower' (Art of Brilliance, 2020). This company not only delivers corporate training for large businesses and employers, but also offers assemblies, classroom sessions and INSET training days for schools, stating that 'resilience is a learned behaviour' (Art of Brilliance, 2020). Similar programmes across the globe have emerged that specifically target children and young people, such as 'Bounce Back!' in Australian schools, also recently piloted in Scotland (Axford, Blyth and Schepens, 2010). A key characteristic of these initiatives is how they individualise the ability to 'bounce back' and couple this 'superpower' with potential future gain.

Damien Hinds exemplified these ideas in his key speech in 2019, defining character as "believing that you can achieve; being able to stick with the task in hand; seeing a link between effort today and payback in the future" (DfE, 2019b: 5). These ideas of future reward and deferred gratification are embedded within the character agenda. They also tie into wider trends geographers have identified concerning young people's transitions in neoliberal times and the role of education, particularly in the United Kingdom, in shaping an 'aspiration nation' (Brown, 2011; Holloway and Pimlott-Wilson, 2011). Indeed, many of the DfE's recent ideas around grit, resilience and bounce-back are coupled with a focus on aspiration. Their recent call for evidence on character and resilience sought views and examples on "believing that you can achieve (e.g. being self-confident, believing in your own abilities)" (DfE, 2019d). This connects to research that has highlighted the growing 'individualisation of success or failure' (Pimlott-Wilson, 2017; Jerome and Kisby, 2019) where (resilient) young people as self-entrepreneurs gather experiences in order to enhance their employability and 'stand out from the crowd' (Holdsworth, 2017). Although many of the virtues and values encouraged through character education are very easy to support, such as tolerance, fairness and kindness, it is worth stressing that other attributes and traits are laden with a specific individualistic meaning in a neoliberal context, for example its increased focus on fostering 'drive', 'focus' and 'motivation' in young people. One can ask, critically, whether these terms are actually about character, and can be considered in the same vein as kindness and fairness? This debate relates to the wider definitional dilemmas of character education outlined in Chapter 2.

The focus on Britain's economy as a driver for character education, introduced in Chapter 3, dovetails here with this focus on individual resilience, grit and perseverance. As Nicky Morgan declared when discussing character education, "In the 21ˢᵗ century there is no such thing as [a] job for life. Many of today's pupils will have up to eight different careers during their working lives" (2017: 80). The focus on bounce-back can therefore be understood as preparing individuals for a precarious employment journey throughout the life course, with Morgan stating it is the business community who need young people to have resilience and enthusiasm in their work. However, it is clear that this framing by UK Government has failed to account for the generational challenges young people face with unsecure and precarious employment as well as debt-related and housing difficulties (Smith and Mills, 2019). Spohrer and Bailey (2020) argue that character education's links to economic growth and social mobility can be understood through biopolitics, with an intensified drive for self-investment underpinned by ideas of human capital. They highlight the implications of this logic whereby those from disadvantaged backgrounds are seen as more 'in need' of such interventions and 'better' self-management. Indeed, much of this section's discussion is laden with ideas about the wider geographies of class, gender and race. Furthermore, it is clear there is also a growing trend for 'bounce-back' and grit to be coupled with well-being and mental health. For example, in the DfE's latest character education framework guidelines, it states that "having good coping skills (part of being able to bounce back) is associated with greater wellbeing" (2019a: 7). As Hinds states, "There have always been stresses and pressures with growing up. But for today's young people there are also new and heightened pressures, partly due to the evolution of technology and media. This is also closely related to character and resilience" (DfE, 2019b). These social constructions of childhood and debates about technology will be returned to later in this chapter.

As hinted above, the lofty ideas surrounding ambition and aspiration in relation to gritty resilience and bounce-back do not tackle, or even hint, at the structural inequalities shaping the lives of children, young people and families (Jerome and Kisby, 2019). For example, how lived experiences, social mobility and life chances are shaped by class, gender and race. Indeed, educator Love (2019a) outlines how grit, for example, is one of a series of 'quick fixes' that "pathologize African-American children and are inherently anti-black" (2019b). They outline how this term and the wider character education movement in the United States function as "educational *Hunger Games* propaganda" (2019a: 73). Her powerful argument is that "Grit is in our DNA" but that "grit alone will not overthrow oppressive systems of power…

We need educators who understand the legacy of African-Americans' grit – that we have survived because of it, and the surviving is not enough" (Love, 2019b). This wider work on what they term the 'educational survival complex' within the United States, which includes character education and 'grit labs', is instrumental in demonstrating how seemingly mundane trends or buzzwords in educational theory harbour much deeper-rooted ideas and sustain racial injustice. Sayer (2020) has observed how debates on inequalities and social mobility in the context of the character education movement mean the whole concept of character is now met with suspicion. They call not only for continued critical engagements with the term in such contexts, but also for a partial 'rescuing' of the term character and its significance, outlining some similarities with Bourdieu's concept of habitus, as well as encouraging sociologists and social theorists to consider it afresh. Indeed, Sayer outlines how "while advocates of character education have ranged from the Hitler Youth Movement to Martin Luther King, the political and economic dominance of the Right has ensured its take on character has had the most influence on policy" (2020: 476).

A recent DfE call for evidence on character education welcomed views on "being able to bounce back from the knocks that life inevitably brings to all of us (resilience)" (DfE, 2019d: 9) and there are two key analytical observations to make here about this statement. First, these universal messages equalise current and future encounters with difficulty, suggesting that the quantity and severity of life's knocks will be something 'all of us' face universally. This erases the existence of poverty, inequality and structural issues of marginalisation and exclusion. Second, this framing places the responsibility to succeed or deal with such setbacks squarely on the shoulders of individual children and young people. This reveals some wider moral geographies of responsibility and citizenship whereby the attainment (or not) of values-based attributes such as grit and bounce-back are inscribed at the individual subject level. As this section's focus on grit and bounce-back has demonstrated, there is a wider question to ask about whether the moral norms that character education proposes (at present) create a further entrenchment of the idea that young people are 'not good enough'. This is a defining feature of neoliberal times. Could, in fact, this 'new' shift towards moral education be demoralising? Indeed, in placing so much focus on the need to develop bounce-back for both economic success and good mental health and well-being, the pressure to learn such ambiguous and subjective virtue-based skills on top of academic attainment is worthy of critical reflection. If one fails to cope with life's 'ups and downs', the landscape outlined in this section suggests that this is increasingly understood by UK Government as one's own individual responsibility and due to a lack of grit and gumption.

To further illustrate the argument around the erasure of socially differentiated experiences and encounters with hardship, this section turns specifically to discuss social class and its place in the character agenda. In Damien

Hinds' key speech on character education, the then Secretary of State for Education stated that:

> Character and resilience are the qualities, the inner resources that we call on to get us through the frustrations and setbacks that are part and parcel of life. How do we instil this in young people, how do we make sure they are ready to make their way in the world as robust and confident individuals?
>
> (Hinds, cited in DfE, 2019b)

This framing of young citizens in relation to confidence reveals wider discourses of social mobility and social class. He acknowledges the existence of 'public school confidence', describing this as a form of "bravery, gumption, maybe even a stubborn determination to succeed" (DfE, 2019b). This masculinist framing is striking, with the gendered dynamics of these ideas over time discussed later in this chapter. Hinds continues by describing the benefits of:

> a kind of 'have a go' assertiveness that you have from certain types of school…this confidence is clearly not something that should be the prerogative of those whose parents are able to give them an expensive education. All children should have it.
>
> (DfE, 2019b)

Although Hinds recognised the unequal geographies of access and educational opportunities in this speech, the admission rightly attracted critical commentary, such as that from McInerney (2019) writing in *The Guardian*:

> children today often develop their resilience in circumstances Hinds ought to be trying to change. Right now, for example, across the country there are more than 100,000 children living in temporary accommodation. That's three times more than in 2010 and directly related to the punitive welfare system brought in by Hinds' Conservative party.

Indeed, the lived realities of austerity policies imposed over the last decade, as highlighted in Chapter 3, cannot be separated from an analysis of this educational policy. McInerney also identifies the 'middle class suppositions' of DfE's foundations of character, including sport, performing and creativity, which Millar (2015) in the same newspaper notes "look exactly like the sort of thing you'd see in a prospectus for Eton".[1] These observations clearly link to wider debates on private schooling, privilege and social class, which have been the focus of academic research (i.e. Brown, 2012; Gamsu, 2018; Sutton Trust and Social Mobility Commission, 2019) and media debate, especially given the dominance of UK Prime Ministers and Parliamentarians from independent schools. This is a strong and pervasive idea within character education circles, with Nicky Morgan stating that "only some of our schools, often in the independent sector, provide the necessary eco system to develop strong character traits" (2017: 40). Similar observations about equality of access are noted by Pimlott-Wilson and Coates (2019) in relation to the use of Forest School environments by better-funded and well-performing schools. The recent attempts of state schools to emulate activities historically associated with fee-paying schools, for example encouraging pupils to join the Army Cadets (discussed in Chapter 3), are therefore infused with class-based imaginative geographies. To return to McInerney's earlier observations, they are keen to stress that grit and determination are evident in other areas of working-class lives, but these do not neatly fit into the DfE model. For example, young people who perform everyday routines of caregiving. They ask "is being responsible for picking up five siblings from school and getting their dinner ready, as some children do, a form of volunteering that would build character, or does it only count if you're wearing a Scout uniform?" (McInerney, 2019). This provocative question reveals the wider class-based politics and geographies of the character education debate. Holloway and Pimlott-Wilson (2014a) identified in their research on extracurricular enrichment activities in England how the desire of working class parents for their children to participate in activities such as music lessons, uniformed youth organisations and drama classes is as strong as middle class parents, but access to these opportunities is shaped by financial resource. Their research identified higher participation rates in these types of activities within middle class households compared to working class households, which then feeds into a wider 'institutionalisation' of childhood more broadly.

The focus on grit and bounce-back within character education circles also highlights the pursuit and challenges of cultural capital, namely "the symbols, ideas, tastes, and preferences that can be strategically used as resources" (Scott, 2004: 142). Friedman and Laurison's (2019) recent work on the 'class ceiling' is important to highlight here on structural inequalities and unwritten 'rules of the game'. Their work on graduate employment shows how the charismatic 'soft' skills of confident presentation are vital for entry into the labour market and success in certain employment sectors. However, this cultural capital is accessed and fostered through class-based development opportunities. Essentially, this reveals that a stubborn, gritty determination to succeed will only get you so far, shattering ideas of a meritocracy. There appears to be a growing realisation of this challenge within DfE policies. For example, the latest DfE framework for character education requires schools in England to reflect on six character benchmarks and assess how well their curriculum and teaching develop resilience and confidence, including the question "does it teach knowledge and *cultural capital* which will open doors and give them [pupils] confidence in wider society?" (DfE, 2019a: 5, emphasis added). The very presence of the term cultural capital here suggests an awareness by DfE of these wider inequalities.

However, quite how individual schools and teachers alone can redress these wider historical and systemic inequalities remains unanswered. Indeed, inequalities in education, employment and wealth will clearly not be solved by merely introducing character initiatives in schools. And yet, some champions of character education and schools that have fully embraced this idea espouse that grit is the trait that can seemingly (and universally) overcome such obstacles. For example, one English school boldly claims that "glass ceilings are not acceptable" (cited in Morgan, 2017: 119). The remaining sections of this chapter further analyse the place of grit and gumption as part of these moral geographies of education, starting with mapping their use over time and charting their re-emergence in the character agenda.

From Victorian ideals to measuring morals

Grit and gumption have often been framed by champions of character education as a fresh contemporary 'tagline'; a new idea beyond resilience that captures current needs. However, a focus on grit essentially returns us to some earlier historical contexts and notably repeats late Victorian discourses of character-based improvement. Taylor (2018) has identified how the language of grit and vigour in the late nineteenth century was connected to moral stigmas surrounding poverty, improvement, and encouraging individual success (see also Sayer, 2020). The historical echoes of these ideas within this chapter's contemporary examples are striking, especially around grit as a vehicle for social mobility. Grit and gumption have also appeared in different historical contexts within England, used to invoke ideas of good (or bad) citizenship within schools. For example, in the discharge of 'bad characters' from schools in 1899[2] or evidence collected on the 'state of public morals' from an inquiry into secondary schools in 1961.[3] The language of grit in these varied educational settings and time-periods demonstrates its longevity, tailored to specific moral panics at different points in time. The parallels between historical and contemporary moral projects have been highlighted in other educational spaces. For example, Bull's (2016) research on a classical music education programme for working class children in the United Kingdom, which they describe as a 'bourgeois social project'. Bull argues that this recent example draws on Victorian ideas of culture as a 'civilising influence' and demonstrates "a middle-class disposition by rewarding investment in a future self" where classical music education "cultivates an ideal of hard work as a moral project" (2016: 120). Similar sentiments are clear in the character education project in England.

Despite the fresh new launch surrounding character education in England in recent years, with flashy awards and plaques for schools, there is an awareness by many champions of its historical antecedents and the Victorian era in particular. For example, Nicky Morgan includes a poster in her book entitled 'character is power' given to her by a Headteacher from Surrey (Morgan, 2017: 28). The image by William Briggs, a Toronto-based publisher and preacher from 1899, has four corners labelled 'zeal', 'pluck', 'plod' and 'nerve'. These labels exhibit clear synergies to the contemporary language of bounce-back and stickability as well as the poster's other virtues-based declarations of 'princely manhood'. As hinted earlier in this chapter, ideas about grit and gumption can also be read through the lens of gender in historical and contemporary contexts. These terms draw upon well-worn ideas of a gritty, masculinist 'hardiness' needed to weather life's storms, coupled with a rugged and unshakeable confidence to drive success. Taylor (2018) has identified how many of these notions emerged from late Victorian ideas of manly 'vigour' and steely nerve. These ideas are further entrenched by the focus in contemporary character education material on chivalrous knights, folklore and examples of rousing masculinist virtues (Jerome and Kisby, 2019). It would be easy to dismiss the place of such images and sentiments in contemporary material and schools as romanticised nostalgia. Instead, it is important to recognise the problematic class-based assumptions and logics inherent in the re-circulation of such images or ideas as they are given 'new life'.

Other long-standing features of school life such as reward systems and school ethos mission statements are re-envisioned and re-fashioned in this context. For example, the UK Government praised a community primary school in Swadlincote, Derbyshire for its 'positive behaviour rewards system' to help children to reach their "ideal selves" (cited in Whittaker, 2016). The idea of behaviour-based awards and rewards in school, competing as individuals or collectively as a form or house group, is not a new idea in the English education system yet is now re-imagined within the character agenda. For Morgan, "even those schools which might profess not to teach character will spend time building routines, implementing reward schemes and working on behaviour including skills such as self-control – in other words, there is always a hidden character curriculum" (2017: 117).

The renaissance of 'school ethos' is worthy of more detailed examination for its communication of ideas about grit and gumption. Introduced in Chapter 3 as a second pathway of approaches in schools, this is an attempt to capture and foster a wider community of shared values. School ethos is an identity-based vision designed to reflect a school's culture, ideally one that supports individual flourishing and the collective hopes and aspirations of that environment (Manchester and Bragg, 2013). These devices are therefore an attempt to define a school's own character externally, whilst simultaneously acting as internal encouragement to improve an individual pupil's character. A school ethos is often captured or translated into a shorter school mission statement or motto, which as Chapter 3 introduced can be visually represented within the built environment. For example, the 2015 Character Education Award winner King's Leadership Academy in Warrington, England has a motto of 'Credimus' – Latin for 'We believe' – as well as displaying the slogan 'excellence is habit' above the entrance hall (Morgan, 2017: 44). A school motto is a very common feature of the English education system, again with a long history and class-based associations. The recent revival of them within the character agenda is noteworthy. Take, for example, Nicky Morgan showcasing an example from a school in the West Midlands where a sheet with potential values was sent home for children, parents, teachers and governors to select and vote for, in order to give ownership over the school's future direction. The final six values selected by the whole school community were "respect, kindness, friendship, perseverance, honesty and responsibility" (Morgan, 2017: 110). The selection of perseverance as a virtue demonstrates the idea of gritty resilience in such contemporary exercises. The co-creation of a shared set of school values is attractive for its participatory nature, yet does tend to paint a rosy picture of school life and teachers' working conditions. For example, teachers' union NASUWT found that "eight out of ten teachers do not feel valued or respected" (cited in Roach, 2016: 24). Relatedly, O'Brien (2020) recently asked whether stated school values that are "often front and centre in a glossy prospectus and website or printed on lanyards to ensure visibility" are anything more than slogans, provocatively asking whether values are 'lived out' by staff or school management and whether "policies and decisions are consistent with those values" (O'Brien, 2020).

School ethos and mottos are a clear example of a collective mechanism utilised to encourage effort by individual children and young people to improve their character. These shared visions – whether adult-imposed or co-created by school communities – are increasingly encouraging pupils to persevere and be resilient. They are a rallying call for pupils to embody grit and gumption, both for the benefit of their own individual futures and the school's success. It is worth stressing that many schools in England have incredibly similar mottos and values-based statements or slogans, calling into question their unique distinctiveness. Nevertheless, a motto or school ethos can be a powerful and inspiring motif for a child or young person, perhaps even embraced as their own moral compass. A school's ethos or values-based motto can also be completely ignored or foster ambivalence, reflecting a range of lived experiences for pupils in primary or secondary schools. For geographer Finn, school ethos is "too easily imagined as collective but free-floating from socio-material and historical circumstances" (2016: 32). His proposal to instead consider 'atmospheres' within the school environment and more fleeting collective and individualising feelings is compelling. Nevertheless, for character education specifically, the labelling of school ethos is a popular mechanism for trying to represent and (re)produce ideal citizens and a school environment that fosters grit and gumption.

In some cases, school ethos and mission statements go a step further in such visions to powerfully shape ideas about 'ideal selves' and future citizen-subjects. The analysis of original material underpinning this chapter reveals some important references to an ideal 'golden child' or 'characterful child'. For example, Nicky Morgan draws on the example of a school in Barnet, London that uses the figure of a 'golden child' to encapsulate several values-based attributes via a poem, including not only being 'kind', 'honest' and 'a good friend' but also resilient', 'ambitious' and 'strong in commitment' (2017: 23). The school "makes it clear that, by the time their pupils reach Year 6, they will have developed [this] list of positive traits which make them a 'Golden Child'" (Morgan, 2017: 22). It is unclear if this poem was created by children themselves as a participatory activity or driven by teaching staff and governors. However, the figurative work of the golden child here is noteworthy as an ideal mode of self, constructed through moral geographies that reproduce ideas about characterful citizen-subjects. It also chimes with wider ideas about the behaviours, skills and 'norms' seen as necessary for good behaviour and a learning subject (on these debates see DCSF, 2007: 4; Gagen, 2015; Holt, Bowlby and Lea, 2017). Overall, the examples in this section of revamped reward systems and the re-fashioning of school ethos demonstrate how ideas to encourage good behaviour and promote gritty resilience have historical roots but contemporary salience in the current character agenda.

One aspect of this debate on grit and gumption that has moved on, and distinguishes the historical echoes discussed in this section from contemporary educational practice, is the attempt to measure morals. Indeed, the current educational system in England is a landscape increasingly dominated by data-based monitoring of educational progress as well as wider technological change (Facer, 2011; Finn, 2016). As with many educational outcomes, there are a range of views over the ability to measure, test or evaluate 'grit' or 'gumption'. For many researchers and educational practitioners, grit is a subjective attribute and thus unmeasurable. Yet for others, this characteristic (and its relationships to passion, perseverance, success or personality traits) can be subject to different forms of analysis and evidence-based testing (see Credé, 2018). Indeed, the US-based non-profit organisation 'Character Lab', whose co-founders include Angela Duckworth, has invented a scoring system for pupils and teachers via 'Character Growth Cards'. Character Lab has a growing network and reach in the United Kingdom and is discussed by Love (2019a) for its role in the wider character education movement in the United States. Nicky Morgan highlights the Lab's scoring programme in her 2017 book, which is an attempt at self-evaluation that includes the quantification of grit, optimism, self-control (school work), self-control (interpersonal), gratitude, social intelligence, curiosity and zest (Character Lab, 2016). These terms, particularly grit and zest, again return us to some of the historical contexts discussed earlier in this section and their classed, gendered and racialised discourses.

The wider debate around measuring morals reveals wider questions about the purpose of education more broadly. For example, the journalist Millar (2015) argued that whilst the advent of character education in England via the Character Innovation Fund positively acknowledged "children cannot live by drilling and exam results alone", there were risks that character would become "prescribed, measured to death, and then have the life squeezed out of it by ministerial meddling". Her prediction has largely borne true, with the cabinet (re)shuffles outlined in Chapter 3 leading to a loss of momentum for character education, and the growing move towards the measurement and inspection of character via Ofsted (DfE, 2019a). The question of whether grit or gumption can or cannot be tested, or if it is an accurate predictor of future success, is not the focus of this chapter (see Credé, 2018). Instead, this analysis of the place of grit over time within the character education agenda helps to map the wider moral geographies of education and relationship between character, citizenship and values.

Extracurricular activities and 'the great outdoors'

This section focuses on the role of extracurricular activities, introduced in Chapter 3 as a third pathway for schools to pursue character education. Its discussion illustrates how outdoor adventurous activities and nature-based learning encounters are understood by DfE and other actors as fostering bounce-back and perseverance. Extracurricular activities are defined as those enrichment experiences offered or perhaps co-ordinated by a school, but that are usually 'hooked' onto or appear beyond the school day; they operate via individual parental subscriptions and in the wider community (see Holloway and Pimlott-Wilson, 2014a). These activities usually include sports clubs, youth groups, learning a musical instrument, or creative activities such as drama classes. Co-curricular activities are school-organised, running alongside lessons, and these types of activities – again largely sport-based or creative in nature – can be delivered by teachers or via partnerships with volunteer-led organisations, charities or military ethos providers. There is overlap between these two definitions, but for the purpose of this section the term extracurricular activities will primarily be used given its prominence in the policy and documentary material as offering confidence boosting activities in the wider context of this chapter's discussion.

The DfE have encouraged schools to engage in extracurricular activities across all five of its foundations of character – Sport, Creativity, Performing, Volunteering & Membership, and World of Work (DfE, 2019b), recognising a number of schools already engaged in these areas through long-standing activities. However, the framing of extracurricular activities has more recently been framed in terms of encouraging gritty resilience. For example, Damien Hinds used the example of sport and the 2019 FIFA Women's World Cup to state "I want every child to have the perseverance, grit and determination of the Lionesses" (Hinds, 2019). He linked the England team's character at this elite level tournament with school sport, stating that "PE is on the national curriculum throughout primary and secondary school for good reason but it shouldn't be the only opportunity to get active – extracurricular clubs help young people build character, confidence and employability through volunteering" (Hinds, 2019).

Individual schools in England that embrace this agenda are highlighted by key champions in UK Government and wider networks as examples of best practice. For example, Morgan commends Gordano school in Bristol for its 'Gordano Guarantee' on extracurricular activities, including access to uniformed cadets, Scouts, and Guides (2017: 103). These types of activities are again infused with class-based imaginative geographies, revealing some long-standing associations between certain types of leisure activity, cultural capital and social mobility. Research on access to enrichment activities inside and outside of school walls continues to reveal stark inequalities (Holloway and Pimlott-Wilson, 2014a; Bull, 2019b) with the Social Mobility Commission recently describing this as an 'unequal playing field' (Donnelly et al., 2019).

In addition to weekly or monthly classes in sport, drama or music as extracurricular enrichment activities, gritty resilience is also seen as achievable by DfE and individual schools via hard-worn experiences in 'The Great Outdoors'. These are perceived as boosting the resilience of children and young people and 'comfort-zone busting', with children and young people "trying out new activities in unfamiliar environments, collaborating with pupils from other schools, working with new people" (DfE, 2014a). There are powerful and long-standing ideas about the character-building role of outdoor activities and environments for children and young people and the transformative 'place' of nature in educational landscapes (Hickman Dunne and Mills, 2019). The role of leisure, sport and outdoor recreation, and the associated values and virtues usually associated with the countryside, have been traced by a number of historians of education (i.e. Freeman, 2011; Edwards, 2018). Indeed, outdoor rural settings have historically been understood as ideal learning environments for citizenship training, drawing on masculinist and able-bodied constructions of the 'good citizen' that are entwined with ideas of Britishness (Mills, 2013).

As urban space became more developed in Western societies during the nineteenth and twentieth centuries, the countryside became more revered and celebrated as its antithesis. Children and young people were seen to be in greatest need of such spaces and the physical and social benefits they offered (Gagen, 2000; Gutman and de Coninck-Smith, 2008). These included improved health and fitness, spiritual restoration, moral fortitude and character development – all attributes apparently there to be gained or gleaned from the freedom and open space of the countryside (see Matless, 1995, 1998). The assertion that nature was good for young people was almost unquestioned, and this belief remains a pervasive theme in contemporary society, continuing to harbour moralistic class-based ideas.

The recent character agenda in England has further cemented these connections, for example the DfE character framework (2019a) includes glowing examples of schools who complete Outward Bound Trust and National Citizen Service residentials. This promotion of outdoor activities as part of school's co- and extracurricular activities further reveals the blurring of boundaries between formal and informal education that is central to this book's argumentation. The use of Outward Bound Trust residentials by schools is instructive in this regard, with Hickman Dunne (2019) demonstrating how schools utilise the established charity's challenge-based courses to boost confidence and resilience. Her research illustrates how children and young people largely enjoy these experiences but also can make some young people feel 'out of place' if they lack certain skills or material items, creating 'norms' associated with young people's bodies and outdoor citizenship.

The final part of this section demonstrates the wider place of nature-based encounters in the character education agenda, specifically through the DfE's 'My Activity Passport' initiative. In November 2018, the DfE published a 'bucket list' of everyday outdoor experiences for primary school children. These optional staged activities such as watching the sunrise and climbing trees were badged by the DfE as building 'character and resilience' and achievements to complete before leaving primary education. Activities such as building rockets and sleeping under canvas were inspired by Hinds' visit to St Werburgh's Primary School in Bristol who had devised a 'passport' of similar activities for pupils (Vaughan, 2018). A more formalised 'My Activity Passport' for schools was launched by DfE in January 2019. This downloadable pack lists 140 activities for primary school children to complete, with 20 activities per year group. For example, a child in Year 1 (aged 5–6) would be encouraged to 'roll down a hill' and 'make a daisy chain' whereas a Year 4 child (aged 8–9) would 'swim outside' and 'create a sculpture trail' (for a full list, see Whittaker, 2018). The linked social media campaign #MyActivityPassport encourages schools and pupils to send records of their activities; however, at the time of writing no data are available on the uptake or evaluation of this DfE campaign. It is striking how similar the ideas in the DfE's 'My Activity Passport' are to the badge programmes of uniformed youth organisations, evoking similar romanticised visions of 'The Great Outdoors'. The DfE initiative is endorsed by the Scouts and National Trust (Whittaker, 2018), with Girlguiding and the #iwill campaign (discussed in Chapter 6) also listed as partners. Indeed, #MyActivityPassport strongly resonates with the National Trust's 2013 campaign "50 things to do before you are 11 and three-quarters". This example therefore further reveals the slippage between initiatives to boost resilience and grit as part of a wider landscape across formal and informal education with multiple actors.

The example of the 'My Activity Passport' also reveals some wider ideas about grit, 'bounce-back' and digital childhoods. A supporter of Hinds' initiative, the Headteacher of King's School Hawford in Worcester, wrote that pupils are "quite often wrapped in cotton wool" (cited in Merrifield, 2018). This popular phrase represents a construction of children as feeble, risk-averse and emasculated, and connects to the ideas of grit and gumption explored in this chapter. Indeed, the recent idea of 'snowflakes' in this context was referenced in Hinds' landmark speech:

> when I go to visit schools, I don't recognise this word snowflakes. I don't recognise that in the young people I meet on my visits. The young people I meet are compassionate, civic minded and hard working ....When I compare you and your peers to who I was at your age, my classmates and I, you have so much more confidence, ambition and gumption than we ever did. But of course we'd expect every generation to be better than the last. What I want is for us to reach higher and wider, to improve further still. To make sure that these opportunities are available for everyone and that we value fully the development of character and resilience in all our young people.
>
> (Hinds, cited in DfE, 2019b)

On the one hand, Hinds commends the current generation in comparison to his own and rebukes the label of snowflakes that suggests fragile 'woke' youth. On the other hand, he evokes this term to justify the need for more focus on confidence and gumption to ensure the nation does not fall fowl to this perceived growing trend. The reference to snowflakes here is a powerful example of how grit and resolve are seen as vital to the future success, and even security, of the nation.

Furthermore, for Hinds and the DfE, the milestones in the 'My Activity Passport' are about "doing stuff that doesn't involve looking at a screen. It's about getting out and about" (cited in Vaughan, 2018). The invocation of 'technology versus nature' here is noteworthy and reflects another contemporary moral panic. A powerful driver for promoting many of these activities discussed in this section is the pervasive belief that authentic encounters with nature are rapidly declining due to social and technological changes, with children's digital activities replacing time previously spent on outdoor (character-building) activities, leading some to fear 'nature deficit disorder' (Louv, 2005). The decline and inequalities in access to children's outdoor play in Britain, and the value of outdoor education, is the focus of much research (i.e. Moss, 2012; McKendrick et al., 2015; Russell and Stenning, 2020). However, a recent paper by Novotný et al. (2020) suggests that heightened and polemic fears about children's lack of time in nature are perhaps ill-founded. Their comparison of historical and contemporary data reveals how children in the twenty-first century have less experience with farming activities than children from 1900, for example, but notably more experiences with nature on fieldtrips and recreation. In addition, fears about children's excessive screen time on digital devices are a moral panic that can be tempered by excellent research on the realities of children's digital lives and complexities of digital parenting (Livingstone and Blum-Ross, 2020). Despite this nuance, fears surrounding time on technology at the expense of exposure to 'authentic' natural encounters continue to gather pace, with a direct reference to screen-time by Hinds in his key speech on character education:

> Of course there is good to be had from these technologies and media as well. But, every hour of screen time is an hour not playing out, not reading, or not sleeping. Time spent making virtual choices is time not the same as making real life choices.
>
> (DfE, 2019b)

The ideas surrounding children's need to reconnect with nature, alongside concerns about the dangers of digital childhoods, remain pervasive. They circulate as part of much wider narratives about the need to tackle perceived social and moral ills of contemporary urban life in Britain. A recent review by the Department of the Environment (Defra) espoused the benefits of nature for making the public 'happier and healthier' and recommended 'every child in Britain spends a night under the stars' (Knapton and Hope, 2019). In a similar argument to those made earlier in this chapter about the lived realities and inequalities of children's lives in Britain, it is worth stressing that many children do not spend a night in a secure, warm home let alone the bonus of a camping trip. Additionally, stars do indeed exist in UK cities. These observations may appear glib, and clearly there are serious issues in this context for example around air pollution in cities that impact children's physical and mental health. Nevertheless, it is vital to recognise the political dynamics at play here and the increased use of sweeping statements about the ability of nature *in and of itself* to address inequalities in contemporary Britain and restore individual and national character.

Conclusion

This chapter has explored the moral geographies of grit and gumption within character education in England. It has examined how the superpower of bounce-back and discourses of social mobility circulate within this landscape, where schools are framed as playing a key role in encouraging more resilient individuals prepared for uncertain economic futures in neoliberal times. Furthermore, the chapter outlined how grit is also framed as tackling life's 'ups and downs', increasingly situated within wider discourses on health and well-being. However, this sentiment erases the structural inequalities that shape the lived experiences of children and young people, particularly in terms of class, race and gender. Indeed, the push for character education in recent years could be viewed as having a demoralising impact and perpetuating the idea that young people are 'not *good* enough'.

The chapter then outlined how grit and gumption are not new ideas, but rather a return to earlier historical contexts, harking back to Victorian values and gender-based ideals. Children and young people in England today are encouraged to develop not only bounce-back and stickability but also the pluck and zeal of yesteryear. Indeed, character education itself has a remarkable 'bouncebackability' with newer 'spins' on similar ideas and practices in school environments. These moral geographies retain under-lying class-based ideas, whether in a school's use of a 'golden child' figure or the gritty resolve embodied in their school's motto. Furthermore, extra-curricular activities such as sport and music are seen as having a key role in creating comfort-zone busting encounters, securing cultural capital, and re-connecting children and young people with nature, with lessons in perse-verance that develop resilience and avoid the label of 'snowflakes'.

In so doing, this chapter has continued to develop the book's central arguments. First, the chapter's critical analysis of the moral geographies of grit and gumption demonstrates that geography matters, tracing how these ideas become cast in space over time. The chapter illustrated how these powerful ideas about children and young people and their place as ideal citizen-subjects 'take shape' in schools and other learning environments, shaped by class, gender and race. The book's second key argument on the growing geopolitical dimensions of character education in England has been less obvious in this chapter. However, its discussion has articulated how a focus on grit and gumption can mirror national (in)securities and concerns about the future resolve and 'hardiness' of its (future) citizenry. These ideas are further developed in the remaining chapters. Finally, this chapter has demonstrated the book's third central argument about the increased blur-ring of formal and informal education. This chapter's analysis of extracur-ricular activities by schools and initiatives such as the My Activity Passport to foster character and resilience reflect the slippage between the spaces and practices of formal and informal learning, a cross-cutting theme across these analytical chapters.

The recent moves to encourage nature-based encounters explored towards the end of this chapter were introduced as part of a longer, enduring legacy of outdoor education in citizenship training. As such, Chapter 5 turns its attention to uniformed youth movements and specifically on The Scouts in the United Kingdom, whose message for children and young people to 'be prepared' has been central to its voluntary activities for over a century. Indeed, grit and gumption have been evoked in scouting texts throughout the organisation's history. For example, the inclusion of this poem, origi-nally from the 1850s, in a key Scout manual:

> For a man 'tis absurd to be one of a herd,
> Needing others to pull him through;
> If he's got the right grit he will do his own bit
> And paddle his own canoe
>
> (cited in Baden-Powell, 1922: 15)

Here, the idea of an adventurous canoe journey is coupled with a mor-alistic and values-based assessment of an individual's 'lifepath', but only if someone has the 'right grit'. The notion of 'paddling your own canoe' resonates with a particular imperial geography of adventure and masculinity as well as notions of landscape and national identity. The next chapter expands on these ideas through a specific focus on The Scouts, which provides a useful lens through which to examine character education within civil society and crafted by voluntary action and efforts, rather than driven by the state. However, this chapter also charts the organisation's growing relationship with DfE and more recent call for young people to 'be resilient' within the wider character agenda, building on this chapter's focus on grit and gumption.

Notes

1. Eton is an elite independent boarding school for boys, based in Berkshire, England. It is well known for famous alumni, including a large number of UK Prime Ministers.
2. The National Archives, HO 45/9959/V29601.
3. The National Archives, PREM 11/3264.

5   Be prepared, be resilient

British youth movements and civil society

Introduction

This chapter examines voluntary spaces in civil society – uniformed youth movements – and the role they play as 'character-building' spaces of informal learning for children and young people. This chapter specifically focuses on the example of The Scouts in the United Kingdom, as part of the global scouting movement, and its moral geographies of citizenship training over the last century. Drawing on data from original archival research, two key examples from the early twentieth century are discussed to illustrate the organisation's values-based ideology and performances in outdoor spaces: first, camping; and second, the 'Scout Farm'. The chapter also demonstrates how ideas of grit and preparedness are encapsulated in the organisation's motto 'Be Prepared', building on the discussion in Chapter 4.

In recent years, this youth organisation's vision has been re-framed as a push to 'Be Resilient', with campaigns over the last decade reflecting the wider character agenda in England explored in the book thus far. This chapter outlines how The Scouts has a growing relationship with Department for Education (DfE) through a number of projects, as well as its wider branding around young people gaining 'skills for life'. In doing so, the discussion further elucidates how the boundaries between formal and informal education are blurring, and how values are increasingly enfolded into wider state objectives within civil society. Overall, this chapter maps the contours of character and citizenship in this youth organisation over time as part of a series of wider arguments on the geographies and geopolitics of education.

British youth movements

The emergence of British youth movements was a curious phenomenon in the late nineteenth and early twentieth century. Perhaps surprisingly, many of these voluntary uniformed organisations are still active today providing structured activities for children and young people living in the United Kingdom. The earliest youth movements were framed along religious lines,

DOI: 10.4324/9780203733066-5

with separate gendered movements for girls and boys. These included, but are not limited to, The Boys' Brigade (1883), The Church Lads' Brigade (1891), The Girls' Brigade (1893), Jewish Lads' Brigade (1895) and The Church Girls' Brigade (1901). As the twentieth century began, further organisations were established open to those of all religious beliefs, including The Boy Scout Association (1908; hereafter 'The Scouts') and The Girl Guides Association (1910), as well as non-religious and co-educational youth movements Urdd Gobaith Cymru (Welsh League of Youth) (1922) and The Woodcraft Folk (1925). These organisations developed a presence at the local, national and sometimes international scale, as popular leisure-based pursuits and spaces of friendship, civic participation and youth citizenship (Springhall, 1977; Edwards, 2018). Although each of these organisations has modernised over the last century, for example shifts towards co-education and other changes in membership criteria, their place in the landscape of childhood, youth and (informal) education in Britain has been remarkably constant. These spaces all have similar core features and activities for children and young people through an informal learning programme delivered by adult volunteers at weekly meetings in age-based sections, as well as regular camps and outdoor activities.

Indeed, nature has played a central role in most organised youth movements across the world – either through direct engagements with outdoor education or through attempts to educate youth about the natural world. Springhall described youth movements as "in effect, a mass leisure outlet for the young working-class adolescent controlled by the largely middle-class, middle-aged, adult" (1977: 125). This importantly highlights the issue of class, yet this argument is simplistic in describing youth movements as mass-leisure outlets. They were not designed as a way of getting youth to access the countryside for occasional, fleeting leisure-based encounters. Instead, youth movements used nature in a number of ways and ascribed various meanings to it as an integral part of their character-building activities and ideas about citizenship. Encounters with nature were important to recruit and retain young people, to gain the support of parents and influential members of society, to aide and structure their learning programmes, and as an arena for their particular brand of outdoor citizenship.

Some uniformed youth movements emphasised engagements with nature more than others. For those formed at the end of the nineteenth century, such as the Boys' Brigade and Jewish Lads' Brigade, countryside and camping experiences were merely a bit part in an overall programme centred on drill, parades and attendance at local Churches or Synagogues (Springhall, Fraser and Hoare, 1983; Kadish, 1995). However, those youth movements established in the early twentieth century positioned nature and the outdoors at the centre of their programmes and activities. For example, the Kibbo Kift Kindred, a precursor to the Woodcraft Folk, was "devoted to strenuous hiking, camping and other woodcraft activities with a strong emphasis on pageantry and ritual" (Prynn, 1983: 81–2). This increased engagement with nature over time can be viewed as a reflection of changing societal attitudes towards the needs of youth and responses to emerging urban concerns and moral panics over working-class youth, introduced in Chapter 4. However, many of the decisions determining the extent to which 'nature' should feature in the programmes of youth movements were based on the political, social or spiritual foundations of each particular organisation and the views of their respective founders and adult volunteers.

More broadly, British youth movements communicated powerful ideas about citizenly duty and encouraged characterful, values-based codes of living for children and young people. These ideas about character, citizenship and values overlapped, reflecting the definitional dilemmas and contested understandings of these key concepts outlined in Chapter 2. Uniformed youth organisations have generally conceived of young people as malleable, and that their training programmes could influence or shape young people to develop habits and citizenly behaviours. These organisations encapsulated and spatialised wider hopes and fears about children, young people, nation-building, security and futurity, positioning their programmes as the remedy and solution to successive 'crisis'. Over time, British youth movements have also reflected ideas about appropriate gendered identities intertwined with this wider, moralistic communitarian project. Indeed, in early twentieth century Britain, "citizenship was ... a central feature of a communal moral project through which individuals subscribed to a vision of shared combative and rousing masculinist virtues" (Freeden, 2003: 277). More broadly, these organisations began to represent certain values in public life and become emblematic of the wider 'state of the nation'.

For example, during the 1950s a small number of communist Boy Scouts were expelled from the movement (Mills, 2011b). In a debate in the House of Lords about these incidents, Viscount Alexander of Hillsborough stated that "the [Scout] organisation illustrates the truth of Emerson's saying about character being the conscience of the nation. I believe that is the real purpose of the Scout Movement".[1] Indeed, the behaviour of Boy Scouts was often seen by commentators, print media and the wider British public as a lens through which to assess the nation (and future nation's) character.

Across the spectrum of uniformed youth organisations in Britain, all professed that they achieved 'good' citizenship as part of a journey. For example, the Jewish Lads' Brigade stated in 1961 that they had "produced innumerable citizens of whom we may be justly proud, and we trust with your support we may continue to produce the first-rate citizens and leaders of tomorrow".[2] This focus on the quantity and quality of (young) citizens is repeated across archival material from all of the British youth movements introduced at the beginning of this section. The Woodcraft Folk, a non-religious radical peace youth movement, expressed similar ideas about the impact of their training and informal education, stating in 1937 that:

> The few hours we have at our disposal each week may seem a small part of the child's life, but we have proved those few hours to be of such tremendous value in aiding the young life to grow 'communally straight' that we can go forward in confidence.[3]

All of these youth organisations, despite their divergent religious or political persuasion, were focused on training citizens and instilling the notion of continual improvement, notably in character-based attributes and particular skills and behaviours, often rewarded through badges and awards. Indeed, this competitive focus of collecting small tokens or textile rewards is a feature of nearly all uniformed youth organisations across the world.

Over the last century, young people's own identities and acts of political agency have shaped or negotiated youth movements. For example, in carving out space for new extensions to their gendered or religious-based membership criteria (Mills, 2011a, 2012). These spaces of informal education shaped, and continue to shape, individual and collective identities and are important sites for socialisation and friendship. Over time, the mission statements, uniforms and wider terminology of these organisations have also been repackaged and rebranded, yet there has been a consistent focus on the motifs of character and citizenship. There has, however, been a shift in reflecting broader extensions to citizenship formations. For example, many uniformed youth movements now encourage members to be 'sustainable' and 'global' citizens. Furthermore, in recent years, many of these spaces have more explicitly 'revived' character within their own messaging. This has (re)established them within the wider state-led character agenda explored in this book thus far, not least to access new funding streams to continue their voluntary work in austere times. These historical and contemporary dynamics in relation to character, citizenship and values are now demonstrated through a detailed focus on one uniformed youth organisation – The Scouts. As the most popular British youth movement of the last century, scouting in the United Kingdom has been a central feature in the landscape of voluntary youth organisations. This example and the moral geographies underpinning the organisation demonstrate the important place and motif of character within civil society and the wider relationship between spaces of informal and formal education.

Scouting: 'improving the standard of our future citizenhood'

The Founder of the Scout Movement, Robert Baden-Powell (hereafter B-P), was seen to exemplify "all that was best in the British character" (Warren, 2004) in the late nineteenth century. Following his role at the siege battle of Mafeking, South Africa in May 1899, his popularity as a colonial celebrity grew amid rising patriotism (Jeal, 1989) and a number of charities such as the Boys' Brigade and YMCA sought his endorsement. B-P's initial ideas and methods of scouting were first discussed in military terms in *Aids to Scouting* (1899) and then developed over a number of years through public lectures and correspondence, emerging as an instruction for individual character training in boys. The rationale for such a scheme hooked into a growing and pervasive fear about the lacklustre physical and moral strength of Britain's youth, viewed as a threat to the nation and empire (Boehmer, 2004; Mills, 2013). Indeed, a strong early motto within scouting, utilising B-P's own initials, was 'Be Prepared'. The role of preparedness, as outlined in Chapter 4, is intended to capture a sense of alertness, readiness and adaptability, which this youth organisation has expressed to its youth membership in different ways over time.

B-P tested his methods in an eight-day experimental camp in August 1907 on Brownsea Island, Dorset, seen by many as the birthplace of the Scout Movement. The mystery of an Island Camp resonated with popular children's fiction and other rousing masculinist imperial adventures (Phillips, 1997). Twenty-two boys were invited to the camp, including ten working-class boys from Poole and Bournemouth's Boys' Brigade and ten sons of B-P's friends who attended public school. The mixed programme was carefully choreographed and structured into daily themes, including lessons from B-P in observation, woodcraft, chivalry, saving life and patriotism. The camp was recalled as a huge success and inspired the publication of *Scouting for Boys: A Handbook for Instruction in Good Citizenship* (1908). This was the catalyst for the Movement's unprecedented popularity at a grassroots level and eventual establishment of organisational structures at the local, national and international scale.

B-P stressed that "the function of Scout training is to develop character in all aspects – physical, mental, moral and spiritual".[4] Youth citizenship was employed and enacted in the Scout Movement through a model of three core values – 'duty to self', 'duty to others' and 'duty to God' – encapsulated in the Scout Promise and Law (Mills, 2013). The Scout Promise – "On my Honour, I Promise to do my best, To do my Duty to God and the Queen, To help other people, And to keep the Scout Law." – has had a number of iterations over the decades and in different international contexts, but has been a consistent feature of the organisation and serves as a moral compass. More broadly, The Scouts operated through a moral landscape of 'good' and 'bad' behaviour. This was understood and regulated both at the scale of the individual young person but also the wider organisation and its collective body of youth. For example, the individual declaration and commitment of promising to keep the Scout Law, a series of character-based statements including "A Scout's Honour is to be Trusted" and "A Scout is Courteous" (Baden-Powell, 2004 [1908]: 44). The symbolism of these ideas extended to the organisation and nation's youth more broadly. For example, the moral behaviour of Boy Scouts was often used as a national barometer to gauge the character of its youthful population.

In the early twentieth century, The Boy Scout became emblematic of the good, law-abiding, young citizen. An ideal, model 'citizen-scout' was used to communicate the ambitions of its project, echoing ideas discussed in Chapter 4 on character education in schools. These ideas were then communicated to Scouts via on-the-ground volunteer instructors, as well as through various objects and ephemera that inculcated a sense of tradition and 'culture of scouting' (Mills, 2013). Adult volunteers were described by B-P as needing strong personal characters and should be individuals who appreciated the moral aims of the movement. Indeed, the performance of Scouters – as adult instructors – was understood as directly responsible for the behaviour and future directions of its captive youth audience. Badges and tests also provided an opportunity to communicate ideas about appropriate behaviours, situated within state-level expectations and wider codes of living for British youth. For example, the organisation stated in 1963 that they:

> will not hesitate to refuse the award of the First Class badge to a Scout ... who is known to have shown lack of courtesy or good manners, or has camped so badly as to impair our reputation as a Movement for good camping, or has evidently infringed the Highway Code or Country Code.[5]

This quote demonstrates the intermeshing of various codes of living for youth and the governing of behaviour to instil good character over time.

Despite its moral framework, scouting was a movement that "prized practise over theory" (Boehmer, 2004: xxvi). Indeed, B-P believed that it was through the very practice of *doing* scouting that one could perform and 'become' a good young scouting citizen in the present, not just as future adults. For example, he encouraged neighbourly acts of active citizenship and at times gave support to specific fundraising campaigns, ideas later encapsulated after B-P's death in the activities of 'Bob-a-Job' week (Mills, 2015). Indeed, in earlier writings, B-P stressed that '*Passive* citizenship is not enough. Only *active* citizenship will do' (1944 [1919], 16–18, original emphasis).

The everyday activities and badgework of Scout Groups in meetings were key to the regular, informal citizenship training of Scouts. However, it was nature that was seen by B-P and scouting authorities as the most powerful transformative tool. They maintained that a young person's character could be renewed and transformed simply through being in the 'open-air'. B-P, as Chief Scout, believed that a Boy Scouts' sense of self could be developed in the countryside and that "the out of doors ... serves as the best medium for self-education in character".[6] The next two sections explore detailed examples of this organisation's engagement with 'The Great Outdoors' and its character-building activities fuelled by scouting ideology. First, an analysis of camping is presented to demonstrate its symbolic role within romanticised 'natural' childhoods and the moral geographies of these regulated spaces; and second, the Scout Farm is discussed to illustrate how good character was encouraged through an agricultural skills-based programme.

Camping

The Scout Movement strongly subscribed to the idea that outdoor activities – especially camping – encouraged and generated senses of citizenship. Indeed, rural sites and settings, and in particular the campsite, have been instrumental in creating, nurturing and celebrating the scouting 'citizen'. Camping, described by B-P as "the great point in scouting" (2004, [1908]: 109), was revered as the purest form of scoutcraft and used to introduce and reinforce knowledge or skills directly related to one or more aspects of scouting ideology, including duty to self, others and God.

There are some early indications from B-P in scouting literature about the central role that nature would play in the organisation, for example his emphatic assertion that the "romance of Woodcraft and Nature Lore" is "the key that unlocks the spirit of the Movement" (1944 [1919]: 20). The centrality of nature in scouting can also be read as a metaphor for the organisation's own naturalness. B-P describes the movement as a "natural growth, not an artificially organised thing" and declared how:

> The outstanding feature about scouting is its naturalness. Nature Study is its key activity. All that is natural in the boy is utilised in our training, the artificial being eliminated as far as possible.[7]

The movement, its methods and the individual boy were all envisioned as a natural and 'pure' space, drawing on the social constructions of rural childhoods as pure and authentic, discussed in Chapter 4. This attempt to eliminate the 'artificial' in the above quote also reveals a purifying and moralistic approach to the informal education of children and young people through this organisation.

B-P's own adventures in the outdoors are often cited in Scout histories: from his childhood escapes into the woods neighbouring the independent boarding school Charterhouse to his backwoods survival skills in the Boer War. These experiences were translated into *Scouting for Boys* where B-P used his passion for nature-study and the outdoors to entice youth to his brand of woodcraft. Indeed, nature (especially camping) was a successful recruitment and retention method used to excite and capture the imaginations of young people. *Scouting for Boys* was filled with stories of men of the nation and empire, whether King Arthur in the forests of Medieval Britain, or B-P himself in Africa. These military-inspired tales in *Scouting for Boys* communicated geopolitical ideas, with youth being called on to 'do their duty' in perhaps unfamiliar outdoor and potentially dangerous landscapes of conflict. Employing these types of geographical imagination fits into the wider imperialistic project of scouting (Proctor, 2002; Alexander, 2009). Geographers have studied how metaphors of nature have been used in the context of colonialism, where 'primitive' and 'wild' individuals were constructed as needing to be 'tamed' and justified colonial violence (Pratt, 1992; Ploszajska, 2000). The Scout Movement hosted views akin to class-based domestic improvement, embodying the idea that some people needed improving more than others. Ultimately, nature-based outdoor experiences were part of its wider citizenship training in order to create domestic citizens with particular roles in this wider British imperial project and the reproduction of colonial hegemony in this geopolitical landscape (Phillips, 1997; Boehmer, 2004; Mills, 2013).

Following Brownsea Island and the publication of *Scouting for Boys*, camping quickly became the main activity of the grassroots Scout troops that sprung-up across Britain. This took a variety of formats: weekend patrol camps, overnight camps during hikes and 'survival' camps. The focus of the scouting year however was Summer Camp – a week or fortnight long immersion in nature. Whatever the format or duration of the camp, Scouts would often sleep under canvas in patrols on the muddy fields belonging to farmers and other land-owners, carrying their own equipment and constructing their own campsites. Each activity or chore on camp carried out by an individual boy was communicated through its usefulness to the camp as a whole unit and as part of their broader character training. On Brownsea Island, the patrol system was used so that each Scout had a role or job to do at a particular time and at a particular place, but that they must 'be prepared' for a call to help elsewhere. This example has often been extended in scouting literature to boys knowing their role or place within British society, as an individual and part of a collective whole, yet they must also 'be prepared' to help elsewhere, notably abroad, for the nation and empire.

Campsite spaces were regulated and governed by the organisation; however they also functioned as a space of mundane and everyday activities, fun and friendship. The organisation quickly projected its vision of an 'ideal' camp to members, which contained wider messages for Boy Scouts about their behaviour, values and character. The movement envisioned the ideal camp in the same way it envisioned the ideal individual Scout: clean, healthy, smart, ordered and regulated. Campsites are a deliberate manipulation of rural space, usually woods and fields, through aspects of urban architectural design such as boundaries, entrances and exits. In his study of the Boys' Brigade, Kyle discusses what they see as 'camping's geographies' where a "reconfiguration of outdoor space is important because it can control what camp is designed to achieve" (2007: 251; see also Cupers, 2008; Dunkley, 2009). The model Scout camp was envisioned as a bounded space that housed well-behaved ordered campers and expelled unclean bodies, dirt, disease and 'undesirable' characters. The organisation viewed the layout of the campsite as crucial in this respect, and attributes of spatial design, position and climatology were included in badgework before camp. Once a site had been selected, a Scout camp's internal layout required separate spaces for food preparation, cooking, eating, first-aid, sleeping and sanitary arrangements. Spaces which dealt with food or bodily waste tended to be located on the margins of camp, as far as possible from the main sleeping and cooking areas. The campfire – a symbol of friendship, warmth and light – was usually located in a special wooded or secluded area to create a magical and spiritual setting, which fostered values-based ideals:

> The purpose of the Camp Fire from the Scoutmasters' point of view must be to try to train boys. That is, to make them better citizens by implanting in them values that will be permanent in that they will remain in them long after they have left active Scouting, and have gone their several ways in the diverse walks of life that they choose.
>
> (Hazlewood and Thurman, 1950: 87)

B-P warned about "unhealthy sites, polluted water, insanitary arrangements, flies, amateur cooking and verdigreased cooking-pots".[8] There were therefore lessons for Scouts in how to order and maintain a healthy campsite, encapsulated in the 'Campers' badge. Disease and health were seen as not only a danger to the campsite as a whole, and to other scouts, but were also a threat to one's own ability to be a 'good' scout. This feeds into much broader moral discourses as a fit and clean body were equated with a fit and clean manliness, nation and empire (Phillips, 1997; Boehmer, 2004). Indeed, in 1924, B-P stated that "in the Scout and Girl Guide movements we are making good progress in the training of clean-minded citizens".[9]

Conduct whilst camping was seen as intrinsic to the reputation of the Movement. B-P stated that "Remember when camping or hiking that a good deal is expected of Boy Scouts. You have to keep up the good name of the Movement." (Baden-Powell, 1936: 91). It was also one way in which ideas about governance and behaviour were communicated to members. By constructing norms and codifying good behaviour, this created a moral geography of membership, akin to that of Matless' (1997) good citizen 'versus' the (usually urban) anti-citizen discussed in Chapter 2. Camp was therefore seen as a way of exposing, disciplining and correcting bad behaviours of Boy Scouts. This was often motivated by complaints from the public or media about Scout camps. One report in the *Daily Despatch* in 1922 criticised Boy Scouts on camp, and in particular their cleanliness, asserting that "Boy Scouts are getting an unenviable reputation for dirty camping."[10] B-P responded that the Group should be named "to give them the chance of being taught how to improve and possibly to save their health"[11], echoing themes of class-based domestic improvement.

A system of reporting was also used to collect data for Headquarters about the camping practices of Scout Groups across the United Kingdom, a method for 'rounding up slack campers'[12] These reports illustrate the regulation and governance with which the organisation came to view the space of the campsite. However, these reports also turned into somewhat unexpected lengthy narrative descriptions of camps:

> The routine was our usual and again worked splendidly, giving us tons of time for scouting practices on top of a jolly lot of "routine" scouting .... Cricket and rounders on the beach, and a tea which made us so strong that we broke the tug-of-war rope, and all together a very jolly time indeed. Camp fires are quite indescribable – as they are unforgettable. On the Sunday, the Revd. Sawbridge was good enough to take our Scout's Own [service] ... a few evenings later we met his challenge for 'Charades' in another jolly evening at Castle Farm .... Space prevents description, if one could, of a camp so packed with incident. The launch of the lifeboat; Callis's machine gun; the Night Watch; killing the viper; the Treasure Hunt and the old smugglers clue; Sully's mumps; the jolly game of kick tin; Flag raiding; and King of the Castle ... real good scouting.[13]

This account from Hackney Scout District of their 1922 Summer Camp at Sea Palling reveals its excitement, fun and games. It also provides an alternative reading on what 'good scouting' is, or is perceived to be, and provides an insight into the campsite as a space of enjoyment, memories and emotional bonding. Indeed, camp is commonly viewed by outdoor education practitioners and volunteers in contemporary contexts as an enjoyable way to bond and unite members of a group and strengthen friendships, as well as creating distance from urban life or digital technologies, discussed in Chapter 4.

Overall, camping was not a homogenous experience for Boy Scouts and ideas from adult leaders and young people themselves meant that inevitably the space of the campsite was influenced by the outside world. Camping was also a place for subverting and challenging the middle-class ideals of adults, "a topsy-turvy world where connections of class, gender and even race could be manipulated, at least for a fortnight or a weekend ... [camp] became a site for rule-bending within increasingly bureaucratized organisations" (Proctor, 2002: 77). Camping does tend to be reified in scout folklore and ideology, with other more everyday spaces such as scouting in cities and suburbs usually overlooked (Mills, 2014). However, the organisation's ideas about character also filtered into other projects and, at times, more overtly moralistic projects, such as The Scout Farm.

The Scout Farm

The example of the Scout Farm in the early twentieth century is less well known than camping but demonstrates how nature was used to communicate scouting philosophy and its powerful moral geographies. This short-lived scheme reflects how the three tenets of B-P's values-based model introduced earlier in this chapter were embodied in scouting practice, and in this case, a full-time educational and domestic programme.

The Scout Farm was run by the organisation between 1911 and 1917 for several boys over the age of 15. This demonstrates the movement's conviction in urban-rural transformations that would renew the moral integrity of individual boys as well as assisting the broader national 'back to the land' movement (Proctor, 2002). The scheme was launched following an article by B-P in *Pearson's Magazine* in 1911 on the poor state of British agriculture in which he stated "give my Scouts a chance of being trained at a proper farm school and see what kind of farmers, bailiffs or farm stewards we turn out".[14] In response to this challenge, The Chief Scout was subsequently offered an unoccupied farm near Wadhurst, East Sussex by Benjamin Newgass, a wealthy retired businessman and former head of the Lehman cotton operation in New Orleans. Buckhurst Farm, and the adjacent Buckhurst Place, became officially known as 'The Scout Farm'.

The Scout Farm ran a two-year vocational training course for boys from London and other urban areas. A place at the Scout Farm cost parents 20 pounds a year, though poorer boys were encouraged through scholarships and fundraising. The farm drew on alternative educational theories at that time and B-P explained that:

> teaching of boys' hands should be just as much a part of their education as teaching their heads ... it is not reading, writing and arithmetic inside school walls which had brought the nation along, but *individual character*.[15]

This focus on character is striking, reflecting deeper philosophical views on the purpose of education and the wider relationship between character, citizenship and values. There are similar contemporary debates about the practical and vocational benefits of alternative education spaces, for example the growing popularity of Care Farms in the United Kingdom (Kraftl, 2013), especially as alternative provision (AP) discussed in Chapter 3. Back in the early twentieth century, The Scout Farm was described as a "national need and a remedy" and that "from the social, commercial and Imperial points of view the need of getting men 'back to the land' in Britain is very great".[16] The realities of the Scout Farm scheme were, however, more complex.

Boys were trained in a variety of farming techniques and skills including dairy, poultry and pig farming. Further practical instructions consisted of cooking, laundry, household chores and account-keeping as part of an overall "sound physical and moral training".[17] This practical instruction was coupled with a sustained focus on health and physical well-being, particularly in reference to the 'open-air' setting of Buckhurst and the 'healthy conditions' of the boys' residence.[18] The daily schedule of life on the farm included exercise with mandatory and voluntary times for gym, as well as strict meal times and physical drill. B-P described that "The Buckhurst boys ... grew several inches every quarter and learnt to look after themselves".[19]

Boys at the farm automatically became Scouts and wore Scout uniform. Their 'school outfit' included farm boots, gymnasium shoes, a brush and comb, Scout Hat and Bible.[20] Scouting was therefore the moral and religious framework that would assist them in their training and enable them to develop a sense of self and character, part of which was to regulate and govern their own behaviour. For example, the Scout Patrol System was employed on the farm in order to "raise [an individual] to a higher standard morally".[21] It is worth emphasising that statements about the patrol system in relation to 'regular' scouting were never this explicit in relation to higher moral standards but are notably more overt in material on the Scout Farm. Perhaps surprisingly though, democratic education was also part of its infrastructure, with one report stating, "they managed all their own affairs, cooked their own 'grub', looked after their own cattle, and had their own 'Mayor' and 'Town Council'". [22]

In 1914, The Scouts boasted that that the Scout Farm had "turned out no failures yet" and indeed many of the boys who graduated from the programme went on to become farmers in Britain as well as Canada, Australia and New Zealand. There were, however, difficulties with administration and funding. Staff began to resign, in part due to the behaviour of the boys. One warden reported that "there is not one boy in eight on average who is really capable of exercising the necessary leadership required for the Patrol System".[23] However, it was the outbreak of World War One (1914–1918) that created the most disruption, with many boys called up to serve. By 1917, the scheme was cancelled, and the farm sold. This example, although short-lived, illustrates how scouting contributed to a wider social reform programme at this time and how the organisation sought to improve individual character through nature-based encounters, in this case agriculture and 'open air'.

The ideas behind the Scout Farm were mobilised in other, more everyday practices of Scout Groups. Nature study was, and remains, a key part of the scouting programme. In the early twentieth century, B-P felt that nature study had also been neglected by mainstream education, and he viewed scouting as an alternative to what he saw as the failings of the state-system. He exasperatingly wrote in 1919 that "the wonder to me of all wonders is how some teachers have neglected Nature study, this easy and unfailing means of *education*" (Baden-Powell, 1944 [1919]: 41). B-P believed that nature study united Scouts as it:

> has its attraction for the youngster whatever his temperament, and where properly and intelligently utilised can supply education in the four lines of our training: character, health, skill and service: but at the same time it can give required un-denominational and understandable religious basis to them all.[24]

For B-P, nature study involved lessons in the 'wonders', 'spirit' and beauties' of nature, each demonstrative of the personality and attributes of God (Mills, 2012). This reflects the wider role of religious beliefs in spaces of informal education and character training.

Overall, the two examples of camping and the Scout Farm demonstrate how 'the great outdoors' was used to communicate various scouting ideas and ideals. Although this youth organisation utilised rural and urban environments in their programmes to make scouting 'work', its institutional geographies were powerfully centred on nature-based encounters. It is these historical connections and the enduring power of these ideas that shape, for example, the DfE 'My Activity Passport' discussed in the previous chapter, endorsed by The Scouts. This demonstrates a long-standing belief in the transformative power of nature to encourage good character and values, as well as foster gritty resilience.

Skills for life

This section examines the contemporary geographies of education within The Scouts over the last decade, which can be read in the context of the wider renewed character agenda critically examined in this book. Over the last century, The Scouts, along with many British youth movements, has undergone several changes. These have mostly centred on membership criteria and extending their activities to a wider range of children and young people, currently boys and girls of all faiths and none aged 6–25. Social and cultural difference has 'stretched' the Scout Movement over time, reflecting the fractured nature of citizenship and belonging in British society despite the organisation's often singular vision of 'British youth' (Mills, 2011a, 2012). The organisation has modernised its programme, with new badges on emerging hobbies, activities and digital technologies. However, adventurous outdoor activities and camping remain central to its discursive and performative geographies. These have retained their symbolic and pedagogic relevance, with the allure and perceived benefits of outdoor education to foster grit and resilience. This section outlines four contemporary trends in The Scouts that illustrate the role of uniformed youth movements in the wider character agenda, demonstrating how it is not just schools that have (re)turned or (re)aligned to these discourses in the last decade.

First, The Scouts has increasingly utilised themes of character education as part of a number of campaigns, adding 'Be Resilient' to its taglines as well as 'Be Prepared'. The organisation recently published guidance for all parents in the United Kingdom – not just those connected to scouting – to help build a 'resilient younger generation', "based on our 112 years' experience preparing young people for the future" (The Scouts, 2019a). This advice included "have a go at something new (and be prepared to fail); learn and pass on a skill; spend a night away from home; chat with someone different from you; achieve something as part of a team; learn to pick yourself up, start again and bounce back" (The Scouts, 2019a). These ideas echo much of the understandings and landscape captured in Chapter 4 on grit and gumption in formal education and the utilisation of extra-curricular activities by schools. For example, a related social media post on this campaign from The Scouts reads "To help young people cope with these ever-changing times, we're unlocking a Scouts secret: how to develop bouncebackability. #SkillsForLife".[25] This framing is striking, not least in how the organisation stresses its experience for over a century within civil society. The focus on character and themes of grit, determination and resilience are also embodied in Bear Grylls, the Chief Scout of UK Scouting since 2009 and at the time of writing Chief Ambassador of World Scouting. His former military connections and celebrity TV status as a survival expert is the modern equivalent of B-P's spirit of scouting, a relationship and legacy recently under the spotlight with a public focus on the Founder's role in imperial Britain (Sawer, 2020).

Second, the language around preparedness and resilience within the organisation has become more powerfully connected to the lifecourse in the last decade. These ideas were exemplified in The Scouts 2018 television advert – 'Nat's Story' – as part of its 'Skills for Life' campaign.[26] This traces a young woman's journey in the organisation showing her last day as a youth member having thrived on its adventurous activities. She leaves home for University after receiving exam results, fixes her broken-down car, and balances part-time work in a kitchen with late night studying. The viewer sees snapshots of Nat playing sport, administering first aid, cooking, and using other scouting skills in everyday life before beginning a career in healthcare. As an adult leader in scouting, we then strikingly see Nat deal with the grief of losing a parent, summoning the courage to read a eulogy. The advert ends with Nat as a young mother in an inter-racial marriage, dropping her son off for his first Scout meeting at her old Scout Hut as the words 'Ryan's first day' appear on the screen. Through capturing the types of challenges encountered on the lifecourse – the 'ups and downs' discussed in Chapter 4 – this visual representation of a transition to adulthood, alongside and shaped by a youth movement, is striking. It also reflects the re-branded sense of 'being prepared' and 'paddling your own canoe' within contemporary scouting. These ideas are also institutionalised in the organisation's continued yet refreshed focus on 'leadership opportunities' with the evolution of its Young Leaders' Scheme for 14–18-year olds in the last decade. These teenagers can volunteer with younger sections of a Scout Group, encouraging them to continue their scouting journey and shape their own identities and futures (Austin, 2020). In this example, the historical ideas of leadership through the patrol system have been recast in a formal personal and social development scheme pitched as boosting employability.

Third, over the last decade, The Scouts has had an increased focus on 'values' within its messaging and infrastructures. For example, the centrality of 'values' in the additional alternative Scout Promise for atheists or agnostics introduced in 2014. Here, the line 'Duty to God' has been replaced with a pledge 'To uphold our Scout values'. This change, following an extensive consultation, reflects not only broader changes in British society in relation to religion (Mills, 2012) but demonstrates the continued belief that there are somehow distinct and recognisable 'Scout' values, ones that would presumably 'set apart' members of the organisation from other young people. A modernised Scout Law, shortened in length from the original version in *Scouting for Boys* (Mills, 2013), still not only reflects ideas of honour, courage, loyalty and trust, but also emphasises that 'A Scout is Friendly and Considerate'. These moral geographies of the Law and Promise continue to be vocalised, embodied performances in Scout Groups around the United Kingdom, with children and young people declaring a commitment to scouting values. This example therefore has similarities to school mottos and school ethos, discussed in Chapter 4. Indeed, these are individual declarations echoed as part of a wider collective body, representing character-based ideals.

Finally, there have been a growing number of partnerships between The Scouts and DfE under the umbrella of 'character education' in the last decade. Chapter 3 introduced the 'Character by Doing' pilot of scouting in schools via a DfE-funded Character Education Grant (Scott, Reynolds and Cadywould, 2016). Other projects led by The Scouts, such as the 2019 pilot scheme 'Scouts Early Years', have also been partly funded by the DfE in recent years. £600,000 of support for this early-years initiative was secured via the DfE's Voluntary and Community Sector Early Years Disadvantage Grant, expanding the reach of the organisation to 4- and 5-year-olds in 21 units within disadvantaged communities (Preston, 2018; Pascal and Bertram, 2020). This trial has reached around 300 children and underrepresented groups in scouting, with plans to expand at the time of writing. Its format "is based on the Scouts Method, but also informed by the Early Years Foundation Stage (EYFS) Framework, a compulsory standard across England (there are equivalent policies in Wales, Scotland and Northern Ireland) supporting people working with children up to five years old" (The Scouts, 2019b). The piloted models include a family-led version and partner-led version where "nurseries, children's centres and day care settings deliver the defined Scout Early Years programme" (Pascal and Bertram, 2020: 12), again reflecting the blurred boundaries between the sites and settings of formal and informal education. Indeed, The Scouts is increasingly courted, co-opted or collaborating with DfE through these funded partnerships, and this reflects a more fundamental change in how the state engages with the voluntary youth sector and vice versa.

Overall, this section has demonstrated how over the last decade The Scouts has more prominently engaged with ideas of resilience, preparedness and grit as the wider character education agenda has gathered pace in schools. This voluntary organisation's century-old call to 'be prepared' is suddenly 'back in fashion', and it has drawn on this historical legacy to

revitalise its activities and secure its future development in times of austerity. In turn, the organisation and its place within this landscape have then
given credibility to some of the DfE's recent partnerships and initiatives on
character education. This example therefore demonstrates the wider shifting relationship between formal and informal education and the state and
civil society more broadly.

Conclusion

This chapter has highlighted the role of British youth movements in civil
society and their character-building youth programmes for children and
young people centred on outdoor activities. The chapter has demonstrated
how these voluntary spaces beyond school have an important historical and
contemporary place within the wider character education landscape. This
chapter focused on the specific example of The Scouts, drawing on historical
material to demonstrate the powerful moral geographies of scouting ideology and its performance in various environments. In particular, the chapter
analysed how camping and the Scout Farm in the early twentieth century
were used to create governable citizen-subjects and encourage good, clean
and moral citizens. In recognising several changes over the last century, the
chapter then outlined how the organisation's key ideas of citizenship, character and values have remained constant and are now 'back in fashion'. The
chapter traced the renewed focus in scouting on preparedness through the
contemporary language of resilience and bouncebackability'. Furthermore,
it outlined how this voluntary youth organisation has a closer relationship
to DfE and UK Government than at any other time in its history.

This chapter has therefore contributed to the overall book's cross-cutting
arguments. First, its detailed analysis of a uniformed youth organisation
over time has again demonstrated that geography matters in understanding
the character education agenda. The discussion in this chapter on informal learning has illustrated the importance of particular spaces, such as the
campsite and Scout Farm, in the construction and maintenance of moral
geographies and citizenship training. These spaces are utilised to 'build
character' and have distinct institutional geographies that subsequently
shape wider landscapes of childhood and youth. Second, the chapter's
analysis of the national and imperial visions of this organisation demonstrated the geopolitical dimensions of this debate and how character and
values are not neutral terms. For the Scout Movement, the construction of
an ideal Boy Scout was a discrete project influenced by ideas about youth
in early twentieth century Britain, fuelled by national and imperial insecurities, preparing young people to 'do their duty'. Indeed, through using a
broader more inclusive definition of the state that extends outwards from
formal institutions into civil society, we can conceptualise youth movements
as part of a wider state effort and apparatus to connect individuals with
popular values and models of behaviour. Finally, the chapter's mapping of

the contemporary dynamics of this youth organisation supports the book's third core argument on the increasingly blurred boundaries of formal and informal education. Indeed, the creation and reformulation of the scouting citizen is an ongoing project. The organisation continues to espouse a moral framework that positions young people as governable citizen-subjects with a call to 'be resilient' as well as 'be prepared', but its influence now extends into schools. This chapter's discussion of the growing and at times interdependent relationship between The Scouts and DfE also reflects shifts in the wider relationship between the state and civil society in austere times.

Uniformed youth continue to have a powerful and symbolic place within the wider youth sector in the United Kingdom. For example, the recent £5 million Department for Digital, Culture, Media and Sport's 'Uniformed Youth Fund' created thousands of new places in uniformed youth groups in deprived areas of England (DfE, 2018). This expanded the work of 'Youth United' and other funded programmes to support The Scouts, Girlguiding and the uniformed Brigades. Indeed, Davies (2017) calculated that these and other uniformed charities received £70 million between 2012 and 2017 during a period of steep decline in state-funded youth work, local youth clubs and detached provision (see also Davies, 2018). Although uniformed youth organisations are technically non-military spaces that operate in civil society, this support from UK Government is noteworthy and chimes with examples earlier in this book around the allure of military ethos providers within the character agenda.

Over the last decade however, uniformed youth movements have found themselves competing with a new kid on the block. National Citizen Service (NCS) has reached over half a million young people in England and Northern Ireland since 2011 with branded hoodies rather than uniformed khaki. Chapter 6 elucidates the connections between NCS and the wider character agenda, with this new scheme encouraging teenagers to develop 'softer' skills of generosity and kindness to become better neighbours, alongside the gritty resilience and gumption explored thus far. Furthermore, NCS reflects a wider shift in state-level discourses away from youth volunteering towards 'social action'. In describing itself as the 'fastest growing youth movement in over a century', a detailed analysis of NCS further reveals the blurred boundaries between formal and informal learning spaces, as part of this book's wider argumentation and mapping of the moral geographies of education.

Notes

1. Viscount Alexander of Hillsborough, Parliamentary Debates (Hansard) The House of Lords Official Report, Vol. 186, no. 46, Thursday, 11 March 1954, 353.
2. Southampton University Archive, Special Collections/MS223/A827/4/15/ Manchester Jewish Lads' Brigade and Club Annual Report, 1960–1961, Jesse Broad & Co Ltd: Manchester, p. 19.
3. Youth Movement Archive, London School of Economics/Woodcraft Folk/1/2 [1934–1938], Reports and Accounts for 1937, Folk Council, no pagination.

4. The Scouts Heritage Collection (hereafter TSHC), Gilwell Park/TC/24, 'Boy Scouts', a letter by Baden-Powell published in *The Church Times*, 10 October 1924.
5. Newcombe, N. W. 1963. *Setting the First and Second Class Journeys*, Boy Scouts Association, London, p. 27.
6. TSHC/TC/83/1920/*Nature Craft for Scouts*, typed notes by Robert B-P.
7. TSHC/TC/83/1919/*Nature and Naturalness of the Movement*, typed notes by B-P.
8. TSHC/TC/43/Camping 1916–1922, *Campcraft*, unpublished article, Robert Baden-Powell, 1919.
9. TSHC/TC/109/Communists 1924 File, Letter, Robert B-P to Mr. Wilson, General Secretary of the British Empire Union, 17 January 1924.
10. TSHC/TC/432/Criticisms, copy of 'Undisciplined Boy Scouts', *Daily Dispatch*, 24 August 1922.
11. TSHC/TC/432/Criticisms, Letter, Robert B-P to The Editor *Daily Dispatch*, 31 August 1922.
12. TSHC/TC/43/Camping 1916–1922, Letter, D.M. Penrose, Deputy Camp Chief to Scout groups in West Ham District, 5 July 1922.
13. TSHC/TC/43/Camping 1916–1922, Report, 'August Camp 1922 at Sea Palling', F. C. Harris, Hackney District, 1922.
14. TSHC/TC/146/*The Scout Farm, Buckhurst Farm*, page of typed notes, 1.
15. TSHC/TC/146/copy of 'Aiding Buckhurst Scout Farm', *The Courier*, 6 March 1914, emphasis added.
16. TSHC/TC/146/Buckhurst Farm, 'The Scouts' Farm' in 1912 Prospectus, 9.
17. TSHC/TC/146/Buckhurst Place Farm, 1911 Prospectus, 3, original emphasis.
18. TSHC/TC/146/Buckhurst Place Farm, Prospectus 1911, 4.
19. TSHC/TC/146/copy of 'Aiding Buckhurst Scout Farm', *The Courier*, 6 March 1914.
20. TSHC/TC/146/Buckhurst Place Farm, Prospectus 1911, 7.
21. TSHC/TC/146/Buckhurst Farm, 'The Scouts' Farm' in 1912 Prospectus, 9.
22. TSHC/TC/146/copy of 'Aiding Buckhurst Scout Farm', *The Courier*, 6 March 1914.
23. TSHC/TC/146/Buckhurst Place Farm – founder's Papers 1912–1920, Letter and Notes, Percy W. Everett, Deputy Chief Scout to Robert B-P, Chief Scout, 2 October 1912.
24. TSHC/TC/43/Camping 1916–1922, *Nature Study*, typed notes by B-P.
25. The Scouts (2019) Twitter Post @UKScouting, 8 December 2019. Available from: https://twitter.com/UKScouting/status/1203579331760533504.
26. 'Skills for Life – Nat's Story' (The Scouts, 2018). Available from: https://www.youtube.com/watch?v=I0NSyX5Td5g.

6 'The lessons they don't teach in class'?

National Citizen Service and social action

Introduction

In her first week as the newly appointed Secretary of State for the Department of Digital, Culture, Media and Sport (DCMS) in July 2019, Nicky Morgan MP visited a National Citizen Service (NCS) programme in Sileby, Leicestershire. The former Education Secretary stated:

> I am a huge fan of the National Citizen Service. I think the motto of 'Just say yes' is so important. I am a great believer in building character, I think that's what NCS does.[1]

Morgan's visit and comments demonstrate her long-standing belief in character education, discussed in Chapter 3, and support for the UK Government's flagship youth programme. NCS is a state-funded youth volunteering programme for 15- and 17-year-olds living in England and Northern Ireland, described as "the fastest growing youth movement of its kind in the world" (Cameron, 2016). The expansion of NCS has fundamentally (re)shaped youth policy and youth work in the United Kingdom in the last decade, and this chapter uses this example to demonstrate a series of interconnected arguments on the geographies and geopolitics of the character agenda as well as the shifting boundaries of formal and informal education.

NCS centres its understanding of character in terms of civic virtues and social action, specifically generosity, giving and neighbourliness. This chapter examines the wider shift in state-level discourses away from youth volunteering towards social action, defined by the UK Government as "volunteering, giving of money, community action or simple neighbourly acts" (DCMS, 2016; see also Birdwell, Scott and Reynolds, 2015). Indeed, social action is used as a synonym for voluntary action, community participation or ideas of 'service' and has been developed through ideas such as the Giving White Paper (2011) and Civil Society Strategy (2018) (see Bennett et al., 2019). The performance of social action in NCS also runs alongside an encouragement to develop the gritty individual resilience needed for success

DOI: 10.4324/9780203733066-6

discussed in Chapters 4 and 5. Overall, this chapter explores how these ideas about character operate in tandem and through powerful references to transitions to adulthood.

This chapter begins by outlining the emergence of NCS and its place within the wider (re)turn to character over the last decade. It elucidates the relationship between character and citizenship and outlines the moral geographies of values and virtues expressed in NCS as generosity, giving and neighbourliness. The chapter then examines the constructions of adulthood which underpin NCS' claims to 'fast track' young people's futures. Finally, the chapter critically examines NCS' claim it provides "the lessons they don't teach you in class" given its increased links with schools, further illustrating the shifting contours of formal and informal education. Overall, the chapter traces how NCS operates as a modern 'rebranded' formulation of citizenly duty focused on 'social action' and examines its place in this wider policy landscape and young people's lives.

"Say yes to adventure": National Citizen Service

The latest aim of National Citizen Service is that it helps to build "a more cohesive, mobile and engaged society" (NCS Trust, 2019). Since its launch in 2011, the voluntary programme has reached half a million teenagers in England and Northern Ireland. NCS offers a four-week Summer programme, or shorter week-long version in Spring and Autumn, delivered to 15–17-year olds by different regional providers. These providers are a combination of private sector partnerships, charitable youth organisations and social enterprises, who bid to win contractual tenders. This system reflects the wider changing nature of youth work and youth services in austere times (Davies, 2018; de St Croix, 2011) and growing trend of measuring 'impact' in this landscape (de St Croix, 2018). Indeed, this discussion must be seen within the context of the decline of state-funded youth services via local authorities in the United Kingdom in austere times (National Youth Agency, 2019). Overall, NCS Trust state that "by bringing together young people from different backgrounds for a unique shared experience [NCS] helps them to become better individuals, and in turn better citizens" (NCS Trust, 2019).

NCS has a wide range of goals and methods that have developed over the last decade, but it is ultimately a short-term personal and social development programme that includes four phases: first, an adventurous outdoor residential experience; second, a 'life skills' focused indoor residential; third, the opportunity to plan and design a social action project; and finally, a fourth phase to deliver this project and graduate. The policy landscape surrounding youth social action is discussed later in this chapter, but popular examples of NCS social action projects include fundraising for local charities; supporting existing charity work such as gardening or food bank collections, and in some cases, campaigning and activism. For both residential experiences, young people are in large 'waves' from the same region, but the later activities are completed by much smaller teams of young people from the same village, town or city. Indeed, the scheme encourages participation at the local and regional scale as part of a national collective. Participants pay £50 to join an NCS programme, with bursaries available for low-income groups.

The original idea and subsequent genealogy of NCS has been covered in detail elsewhere, drawing on wider social constructions of young people, class, race and religion, as well as different political visions (Mills and Waite, 2018). In brief though, NCS was championed by the Conservative Party prior to the 2010 General Election driven by former Prime Minister David Cameron, who at the time of writing is still the Chair of Patrons at NCS Trust. The programme was later brought to life by the Coalition Government of 2010–2015 and featured heavily as part of the 'Big Society' ideology (Mycock and Tonge, 2011b) and later as part of Theresa May's vision for a 'Shared Society' (HM Government, 2017) within the Conservative Party Government from 2016 onwards. NCS was originally housed in the Department for Education (DfE), but later moved to the Cabinet Office and then the Department of Digital, Culture, Media and Sport (DCMS). DCMS retain oversight of NCS, but responsibility for the management of the scheme is led by the not-for-profit organisation NCS Trust (2016–present). NCS received a Royal Charter in 2017 and has received £1.5 billion of government investment to date (for the latest official evaluation, see DCMS, 2020). Despite critiques from the Local Government Association (2018) and elsewhere (i.e. UNITE, 2014; National Audit Office, 2017), NCS retains strong champions across political parties (i.e. Blunkett, 2016; Jarvis, 2016).

The remainder of this chapter draws on the extensive data-set introduced in Chapter 2 to specifically explore NCS' place within the wider character agenda. The following sections trace the programme's relationship to ideas of citizenship, social action, generosity, giving and neighbourliness. It then critically analyses the understandings of adulthood embedded in NCS that frame these moral geographies, before a discussion of this voluntary scheme's growing links to schools.

NCS, active citizenship and social action

Chapter 2 highlighted how research on the geographies of youth citizenship has interrogated the tensions between rights and responsibilities for citizens 'in the making' and moral geographies that cast some young people as citizens and others as excluded from that social contract. Such state-centric definitions have been challenged by the third sector, where various organisations have defined youth citizenship in their own terms; and yet, their activities are often still tied to the state, for example through discourses of participation or tackling youth violence. Researchers have also demonstrated how young people engage in their own independent 'acts' of citizenship via participation and activism (i.e. Lister et al., 2003; Weller, 2007). An examination of NCS brings these different layers together as a scheme that is keen to highlight young people's own articulations of citizenship, but that has itself been created and developed through wider state policies. Chapter 2 also outlined how citizenship training for young people in a range of different spaces has historically relied on the ideological construction of a good (young) citizen and their behaviour and practices, for example in schools (Pykett, Saward and Schaefer, 2010) or youth movements (Mills, 2013), themes explored elsewhere in this book. For NCS, its particular 'brand' of youth citizenship (Mills and Waite, 2017) is centred on social action, which lies at the heart of its 'compact' between (young) individuals and the state.

Social action is defined by the independent campaign 'Step up to Serve', discussed later, and many other youth sector organisations as "taking practical action in the service of others to create positive change" that leads to a 'double benefit' for both young people and their communities (cited in National Youth Agency, 2020). The focus on social action in NCS frames volunteering as a form of citizenly duty, encouraged in its curriculum through designing and planning a social action project. Indeed, this 'brand' of youth citizenship encourages and legitimises a certain type of citizen-subject (Mills and Waite, 2017) mobilised through the liminal period of youth with its continued emphasis on the 'future'. There is a tension inherent in youth citizenship that poses adulthood as a destination (explored later in this chapter) and therefore presents young people with an emasculated version of citizenship while waiting for 'real thing'. For example, the model of (youth) citizenship offered by NCS and its civic virtues of characterful generosity and giving. Furthermore, NCS frames citizenship as primarily confined to local and national citizenship formations through local acts of social action as part of a national organisation. Ideas of global citizenship are curiously absent (Mills and Waite, 2017), in stark contrast to the frameworks of overseas youth programmes and volunteering gap-year schemes (e.g. Baillie Smith and Laurie, 2011).

The primacy of social action as the key marker of a good young citizen in the NCS model, rather than say voting or democratic engagement, reveals a wider state vision. Although successive governments have encouraged youth volunteering through various schemes and initiatives (Davies, 2017), NCS' focus on certain types of activities is revealing. Indeed, the growth of NCS "represents the encouragement of neoliberal citizen-subjects for a neoliberal state in neoliberal times, encouraging a 'type' of citizen that performs 'safe' and compliant acts of (youth) citizenship" (Mills and Waite, 2017: 72; see also Kennelly and Llewellyn, 2011). Indeed, Davies (2017) has noted how government understandings of social action have "carefully glossed over the term's roots in often radical and oppositional forms of collective activity". A key finding of the research project on NCS introduced in Chapter 2 is that young graduates of the scheme often equated citizenship with volunteering. This was evident not only in the survey analysis and qualitative interviews, but also in the analysis of its wider curriculum, infrastructures and materials. These tended to emphasise the responsibilities of NCS participants (as young citizens) rather than their rights, and as a result, citizenship was often a contested, ambiguous or poorly understood concept by participants, alumni and staff. The achievements of young people 'on' NCS and the benefit of their respective social action projects are not in question here, with inspirational and transformative moments for participants on the programme. However, it is important to highlight how state-level discourses of social action are being utilised to position young people and frame their wider role in society, as part of the broader character agenda. In particular, the values and traits of generosity, giving and neighbourliness emphasised within NCS.

NCS and character: generosity, giving and neighbourliness

The language and discourses surrounding NCS have recently shifted from a brand of youth citizenship (Mills and Waite, 2017) to a more concrete focus on character (Weinberg, 2019). Indeed, NCS is increasingly described in relation to the cultural values that it fosters and promotes. For example, the current NCS quality framework references individual outcomes of 'self-expression, emotional regulation, self-efficacy, confidence, persistence, empathy and team-working'. This section outlines how these attributes and other character traits are spatialised within the NCS programme, including gritty resilience as well as its core focus on generosity, giving and neighbourliness.

NCS participants are encouraged to 'say yes to adventure' during its Week One outdoor-based residential. These individual challenges and team-building activities are heavily promoted in marketing material and seen as a key bonding phase of the overall programme that develops gritty determination and 'bounce-back', discussed in Chapters 4 and 5. A range of documentary material on NCS features the word 'challenge' and seeks to push young people out of their 'comfort zones' through adventurous outdoor activities such as canoeing, abseiling, hiking and raft-building. The usually remote spaces of outdoor education centres or campsites in Week 1 are viewed as essential to these learning activities and atmosphere. For example, the group of NCS participants in the research project's ethnographic fieldwork had no access to electricity during the first residential, noting in their animated whiteboard diary that "the camping was horrible and the lack of electricity was hell, but it forced us to bond more as a group and we made 'The Squad'".[2] It is significant to note that whilst these new friendships were initiated in a space where young people were out of their 'comfort zones' and where there was no access to technology, new friendships were subsequently sustained using social media platforms such as Facebook and WhatsApp groups in later stages of the NCS programme.

In contrast to this gritty adventurous risk-taking at the start of NCS, softer skills and performances of generosity and giving are seen as the key character-based outcomes of the social action phases in Weeks 3 and 4. Here, teenagers deliver a local community-based project to benefit the wider neighbourhood as well as their own personal and social development. The social action project is designed to help NCS participants 'change the world around them', and this is what participants in the study understood 'learning citizenship' to be, with 89% of survey respondents reporting they felt they made a difference through social action. In reference to citizenship, young people tended to refer to making a difference in their local community and belonging, for instance one interviewee stated "So I think NCS like taught about, you know, how to be a citizen and what it is to like give back to your community".[3] Likewise, another NCS graduate conveyed how his understanding of citizenship was focused on his local area:

> I learnt about relationships between people and like community areas that I didn't know about, it opened my mind quite a bit about the kind of area I was in, in some ways negatively, but in some ways positively, which is ultimately good.[4]

Furthermore, the research revealed that many social action projects observed or discussed as part of this dataset had a fundraising element. This was either through social action projects that aimed to fundraise for local charities or fundraising activities that were required to buy materials to deliver the social action project. In many cases, NCS encouraged young people to collect donations, sponsorship or resources from family and friends. Whilst NCS has a fixed cost of no more than £50 that can be covered by a bursary if necessary, the hidden costs of social action were more diffuse and unexpected. Indeed, it is important to recognise that not all parents or families have the resource(s), time or opportunity to contribute in various ways that are often assumed or expected, for example to bake a cake, take a sponsorship form to work, or travel expenditure. The money raised for charity nationally by NCS participants is noteworthy, with inspirational achievements by young people including over 12 million hours of voluntary action (Inspira, 2018). However, one of the recommendations of this research project was that stakeholders should be more sensitive about who shoulders the burden of fundraising during social action projects and other hidden costs of social action (Mills and Waite, 2018).

The local acts of social action by young people are viewed in the NCS infrastructure and policy landscape as having a wider societal impact linked to the programme's overarching aims. The notion of a good local 'neighbour' as part of its citizenship model is connected to wider ideas of social cohesion and social mix as a state-led attempt to encounter 'difference' (Mills and Waite, 2018). These ideas also draw on historical connections between responsibility, national duty and service learning (Mohan,

1994; Edmonson, Tatman and Slate, 2009) with NCS' own name evoking that of national service and military conscription (Mills and Waite, 2017). Indeed, one of the original drafts of a curriculum for this school-leaver programme included reference to a 'military week' (Conservatives, 2007). Although NCS is voluntary, one of David Cameron's early announcements about NCS stated:

> I want to see a programme which engages young people and gives them a sense of purpose, optimism and belonging. Something like National Service. Not military, not compulsory, but universal and in the same spirit. It's going to teach them what it means to be socially responsible by asking them to serve their communities ... we must all come together to do more about the national scandal of all this wasted promise. We owe it to the next generations.
>
> (cited in The Telegraph, 2010)

More broadly, NCS and individual citizens have been positioned as the solution to fixing 'Broken Britain' (Mills and Waite, 2018). The pathway to this improved society is by fostering more active, engaged citizens and moulding these into good, generous future neighbours. The belief here embodied in Cameron's Big Society ideology was that neighbours 'look out for each other', potentially reducing the burden on state welfare provision. One of the interesting moral and political discourses running through these ideas of neighbours of 'good character' is the links to friendship and care (Bowlby, 2011; Painter, 2012). More broadly, some studies of character education have focused on ubuntu and the practices of ubuntu-inspired communities from across sub-Saharan Africa with its focus on notions of humanity, caring, sharing and social interdependence (Etieyibo, 2017). However, ubuntu has a contested place within citizenship education in different national contexts (Enslin and Horsthemke, 2004; Letseka, 2012). In England, the character agenda in schools continues to make connections and inferences to 'neighbourliness'. For example, Damien Hinds stated in 2019 that the re-introduced Character Awards for schools would:

> celebrate school programmes that develop a wide range of character traits including conscientiousness, drive and perseverance; virtues like neighbourliness and actions like service to your community, where even something small can have a huge impact on people who live there.
>
> (DfE, 2019b)

Overall, we can critically question the state-led pushes for (certain) individuals to be 'better' neighbours and citizens, and the wider moral politics of generosity, giving and neighbourliness in this context. Young people have historically been framed by the state as the ones with the responsibility to 'fix' society and re-stitch the ties of citizenship back together, an idea that underlies much of the rhetoric surrounding NCS over the last decade. On the one hand, there is a focus in NCS on the collective service of a wider generation of teenagers, working together as neighbours and citizens who are generous in spirit and character. On the other hand, the onus within NCS is still very much on the individual young person to pro-actively join this voluntary scheme, to craft their own social action journey and carry this forward into adulthood.

"Fast track your future": NCS as a 'rite of passage' or 'microwaving adulthood'?

This section interrogates some more provocative theoretical questions surrounding adulthood as part of this chapter's wider discussion on NCS and its values-based moral geographies. It outlines how the transition to adulthood (and citizenship) is manufactured, institutionalised and measured; in this case, through a government scheme that aims over just four weeks to simultaneously facilitate both of these transitions. Indeed, "NCS represents a sustained attempt by the Conservative government to 'couple' citizenship and adulthood together as dual goals for young people living in Britain today to reach 'successfully'" (Mills and Waite, 2017: 73). On several occasions, Prime Minister David Cameron referred to NCS as a 'rite of passage', for example outlining how:

> In many societies, there is a rite of passage marking the moment that young people turn into adults, taking on board new rights and new responsibilities.
>
> (Cameron, cited in NCS, 2012)

Here, adulthood is envisioned as a destination to which one must journey. Furthermore, and formalised through NCS as a citizen-building programme, is that this 'event' apparently takes place at ages 15, 16 or 17. This book has indicated how young people in the United Kingdom and beyond are navigating a shifting landscape of experiences and potential achievements to shape their imagined futures and 'stand out from the crowd' (Holdsworth, 2017), or simply to try and stay afloat, in an era of neoliberal governance (Pimlott-Wilson, 2017). NCS' focus on 'life skills' chimes with this wider search to boost CVs and navigate uncertain economic times. Two of NCS' slogans – 'create your own future' and 'fast track your future' – therefore illustrate a tension about the limits of young people's agency to genuinely create one's own future.

Interviews with key stakeholders in this research revealed that the initial working group for NCS spent time reflecting on its appropriate target age range. An original architect of the programme recalls "We didn't see this as 'microwaving adulthood', as someone once said, unkindly. It's like we've seen it as part of a journey towards adulthood".[5] Young people's transitions to adulthood across the Global North and Global South are geographical specific and historically contingent (Johnson-Hanks, 2002; Valentine, 2003; Langevang, 2007; Worth, 2009; Day and Evans, 2015). The need for such a scheme in the United Kingdom specifically was understood by another key architect of NCS in relation to the changing social and economic fabric of society where people are starting work later, and that:

> the transition to adulthood is more often marked by getting really drunk and being sick in the toilet of a restaurant or something … so we wanted to establish something positive that was a transition, or the start of transition to adulthood.[6]

Although this might seem a light-hearted reference to getting drunk as a rite of passage, NCS is underpinned by a number of ideas about young people's behaviours and perceived lack of character (Mills and Waite, 2017). It is important to recognise that underlying a scheme that talks about the positive and optimistic attributes of future citizenry, there are some long-standing concerns about young people as troublesome, risky and apathetic.

A further justification for NCS targeting 16- and 17-year-olds was the belief that this is a significant time in a young person's life for making decisions, with lots of rhetoric about skills and CV building that equate successful adulthood with a transition to further education and entry to the labour market. A key NCS figure revealed:

> I guess it was decided that 16 was … seemed like an age when even though it's not 18, which is more classically the kind of 'you're an adult now', it seemed an age when kind of emotionally and sort of in the view of the community, you know, you're no longer a child but you're not quite an adult yet, and you're in that kind of state of flux when in the next year or two you'll be moving out of your very protective kind of cocoon and you'll have a whole range of choices available to you.[7]

It is interesting to note some of the assumptions made here that soon after 16, "you'll be moving out of your protective cocoon", when clearly not all young people in the United Kingdom have that cocoon to begin with. Furthermore, there is a certain irony here in that this 'whole range of choices' has, under the same Government, been reduced for large numbers of young people. For example, the removal of financial support in England to pursue further education opportunities through Education Maintenance Allowance (Murray, 2010) as well as controversial changes to housing benefit support for those under the age of 21 and wider challenges for young people in the housing sector (Wilkinson and Ortega Alcazar, 2017). Despite this lofty rhetoric about 16 or 17 being *the* time of key decisions, the age for NCS participation was eventually lowered to 15-year-olds, in part because of pressures to meet UK Government targets.

Overall, these ideas about adulthood are related to notions of duty, service and 'giving back' discussed in the previous section. From the first announcements about NCS whilst Leader of the Opposition, David Cameron stated that:

> my vision [for NCS] would harness the talents, commitment and energy of our young people by making their seventeenth year, the bridge between their teenage life and young adulthood, a time for them to develop as individuals by putting something back into society
>
> (Cameron, cited in Conservatives, 2007: 1)

There have been some challenges however in the lived experiences of young people's time on NCS in relation to responsibility and the transition to adulthood. In research interviews, most NCS providers stressed the success of the scheme in terms of personal development and young people taking on responsibility for their social action project. There were also attempts by adult mentors observed in the ethnographic fieldwork to 'hand over' more responsibility to young people as part of NCS' 'rite of passage', such as less wake-up calls and young people as 'mini-mentors' setting the day's task. Ninety per cent of survey respondents in this research 'felt more confident as a result of NCS', with 89% indicating they had developed 'new skills'. However, there were tensions surrounding age and responsibility observed during the ethnographic fieldwork. For example, during the indoor residential week, the young people were responsible for the keys to their own rooms, but only mentors had keys to the front door of each block, prompting one frustrated young person to state "we're old enough to have babies or win the lottery, but not old enough to be trusted with a key!?". This example echoes lots of debates in children's geographies about age, legal classifications and the contradictions of 'boundary crossings' from childhood to adulthood (Valentine, 2003; Skelton, 2010). What is striking about NCS though, is that it was specifically designed as a scheme to facilitate the transition to 'adulthood', and so these tensions were understandably even more frustrating for young people to encounter.

Overall, NCS and its voluntary informal learning programme have 'coupled together' citizenship and adulthood as 'achievements' to unlock. However, this section's discussion outlines how this elision of citizenship and future adulthood is complicated by tensions surrounding young people's liminal status. The transition to adulthood is increasingly referred to within the wider character education movement and within schools in England. For example, in the latest DfE character education framework guidance, it states that "schools have an important role in the fostering of good mental wellbeing among young people so that they can fulfil their potential at school and are well prepared for adult life" (DfE, 2019a: 4). The final section of this chapter outlines the relationship between NCS and formal education before offering some final reflections on the wider landscape of youth social action in the context of the character agenda.

NCS, schools and youth social action/activism

NCS originally framed itself as the antithesis to school and the provider of lessons it 'couldn't teach'. However, this section demonstrates that despite an attempt to distance itself from the space of school, there are clear and growing links between NCS and formal education. NCS heavily recruits its target audience via schools and has its own curriculum and 'ethos'. The programme is designed to have learning outcomes for each phase and these are connected to certain spaces, many of which are themselves educational sites and settings. For example, campus-based Universities are popular locations for the Week 2 residential experience on 'life skills' and used to inspire young people to consider higher education as part of a successful transition to adulthood (Brown, 2011). These educational connections are embedded in other parts of NCS too, with all young people 'graduating' upon completion and becoming NCS 'alumni'. Originally conceived as a performative citizenship ceremony (Mills and Waite, 2017), NCS graduation is now a lively occasion that celebrates young people's achievements with the presentation of certificates and awards, yet that invokes educational achievement as a performative 'end point'.

There are also increased attempts to promote NCS within statutory formal education. NCS providers have always used school assemblies as a recruitment tool, but it is ultimately the decision of individual schools about the extent to which they promote this government-backed scheme. The introduction of NCS 'Champion schools' shortly after the scheme launched sought to market the benefits of NCS more forcefully to pupils and parents, hoping it could be embraced as an extracurricular activity – the third pathway to character for schools discussed in Chapter 3. Since then, individual NCS providers have, for example, created revision packs and include memes about homework, revision and exams on their official social media channels. More formally, the parliamentary NCS Act 2017 and Royal Charter was a significant moment in the re-shaping of its educational geographies, introducing formal obligations for schools, academies and local councils to promote NCS.

These links were further cemented in official DfE guidelines about NCS, which encouraged schools that although non-statutory, the programme "can support your school's Spiritual, Moral, Social and Cultural studies, and your efforts to promote Fundamental British values" (DfE, 2017b: 7). These closer links to SMSC and FBV, features of the first pathway to character education via the classroom discussed in Chapter 3, can be read as part of a much broader politicisation of values in educational spaces. These DfE guidelines also stress how "NCS is recognised by UCAS[8] and provides practical skills for employment and life" (2017b: 6), invoking aspirations of higher education as another rationale for school and college leaders to champion the scheme. Recommendations for schools to promote NCS from DfE include assigning an NCS teacher representative, getting to know your NCS provider, and strikingly "embedding NCS in your citizenship and PSHE education" (DfE, 2017b: 12). The guidelines state that "some schools invite their NCS provider to support the delivery of sessions at Key Stage 4" and are involved in "a PSHE or citizenship collapsed timetable day" (DfE, 2017b: 12). These examples again demonstrate the increased slippage between formal and informal education spaces and practices as part of the wider character agenda over the last decade.

The final part of this section's discussion outlines the wider landscape of youth social action that NCS sits within and its growing place within character education in schools. NCS sits within a broader infrastructure of state and civil society support for youth social action, namely 'Step up to Serve' (SUTS) and the #iwill campaign. SUTS was launched as an independent campaign with cross-party support in November 2013 and has gained pledges from over 800 organisations, including NCS, to provide social action opportunities for young people. During his time as Secretary of State for Education, Hinds declared that the "£40m #iwill fund (jointly funded by DCMS and the National Lottery Community Fund) has attracted 20 match funders who have contributed a further £26.5 million to date, enabling more than 300,000 young people to become involved in social action" (DfE, 2019b). This campaign can be seen as part of the wider rebranding of active citizenship and youth volunteering (Mills and Waite, 2017) whereby young people are encouraged to develop the 'habit' of social action (Lamb, Taylor-Collins and Silvergate, 2019).

There are growing links between NCS, the #iwill campaign, and character education in schools in the United Kingdom. For example, a joint statement was published between the #iwill campaign and the Jubilee Centre for Character and Virtues at the University of Birmingham on how "youth social action is an effective and meaningful way to develop young people's character virtues" (The Jubilee Centre, 2014). This statement sought to foreground the role of social action, stating it is an "important mechanism for young people to develop and express their character while benefiting others" (The Jubilee Centre and Step up to Serve, 2014). Notably, this statement reflects the Jubilee Centre's own virtues-based understanding of the debate, promoting how character education is:

> all explicit and implicit educational activities that help young people develop personal strengths called virtues. Character development – and the importance of virtue – should be viewed as a core element of social action, empowering all young people to develop a clearer sense of their relationships with others, as well as of their own purpose in life. Indeed, a sense of purpose – one that is discovered and not imposed – represents a striking outcome of young people's social action
>
> (The Jubilee Centre and Step Up to Serve, 2014: 2)

The statement goes on to list four types of virtues that social action enhances: first, performance character virtues (resilience, determination and teamwork); second, civic character virtues (service, citizenship and volunteering); third, 'moral character virtues' (honesty, trust and compassion); and finally; intellectual character virtues (curiosity, critical thinking and open-mindedness) (The Jubilee Centre and Step Up to Serve, 2014: 2). These ideas and references to social action have been captured in the ethos of many schools, the second pathway discussed earlier in this book, especially by those who have reframed or repositioned themselves within the character agenda. These civic virtues have also been expanded into formal infrastructures and projects. For example, the £40 million '#iwill Fund' introduced earlier which supports projects such as "what meaningful social action looks like at primary school level" (RSA, 2019). This expansion of social action to younger age groups is noteworthy. For example, The RSA4 project encourages social action projects with Year 4 children aged 8 and 9 years old, far younger than the target age range of NCS explored in this chapter (see also Body, Holman and Hogg, 2017).

The #iwill campaign and NCS are also interesting for how they re-frame or 'soften' particular forms of political participation and activism by young people as social action. For example, the global 'school strikes for climate' movement that grew from Swedish teenager Greta Thunberg's "Skolstrejk för klimatet" has been co-opted within the character education movement. This example of global youth activism, solidarity and protest, framed largely in terms of highly mobilised young people in the Global North (Walker, 2020), has been labelled by some stakeholders as 'youth social action'. For example, the Royal Society for the encouragement of Arts, Manufactures and Commerce (RSA), who sponsor a multi-academy trust and number of schools in the West Midlands, argued that the school strikes demonstrate the need to harness 'teenagency', stating:

> Research from the Jubilee Centre finds that those who start volunteering before the age of ten are more than twice as likely to form a 'habit of service' (a commitment to continuing to volunteer) than if they start at 16-18 years of age. It is not surprising then that Greta first took an interest in tackling climate change when she was just 8 years old.
>
> (RSA, 2019)

This coupling of character education research and discourses of volunteering with Thunberg's activism is stark, demonstrating how social action has been used to recast youth protest into more palatable and softer forms of character-building activities. This example highlights the wider contested landscape between social action and young people's political participation.

Overall, NCS is the centrepiece of a much wider series of attempts by the state and civil society to encourage youth social action. NCS is increasingly promoted in schools as part of its recruitment activities but is also positioned as a pathway to character alongside SMSC and FBV in the classroom. The place of NCS as part of a wider social action journey demonstrates how UK

Government continues to espouse the virtuous benefits of youth volunteering as part of its overall quest to shape and mould a 'character nation'.

Conclusion

Over the last decade, youth social action has been promoted by successive Conservative-led governments in the United Kingdom and has driven the growth of NCS as its central programme for young people. This chapter has demonstrated how NCS promotes a rebranded sense of active citizenship, suffused with ideas of character and the encouragement of generosity, giving and neighbourliness. The place of NCS is vital to consider when mapping the moral geographies of education, especially in its attempts to train young citizens for the full responsibilities of adulthood and to 'fast track' their (uncertain) futures. The chapter has also demonstrated how citizenship and adulthood are coupled together within NCS, yet it has also highlighted how in practice this is complicated by ideas of responsibility and young people's lived experiences of the programme. Finally, this chapter outlined the increased promotion of NCS in schools and its wider place in the character agenda.

This chapter has therefore contributed to the overall book's central arguments. First, its analysis of the spaces and practices of NCS and young people's experiences of the scheme illustrates how geography matters in understanding this youth programme and the wider (re)turn to character. This scheme's formation of youth citizenship at the local and national scale is expressed as youth social action and translated as moral geographies of generosity, giving and neighbourliness. These civic virtues are spatialised within NCS as it seeks to host different types of learning experiences in certain places, with unique roles for the campsite or outdoor education centre in phase 1, the University campus in phase 2, and the community-based social action project in phases 3 and 4, mirroring a mobile journey for young people to support the transition to adulthood. The chapter's discussion of this rite-of-passage and its geographical dimensions contributes to wider understandings of the relationship between citizenship and youth transitions in academic debates. Second, the book's next argument that the geopolitical dimensions of the character agenda are important to unpack has been demonstrated through the analysis of this new and rebranded form of 'national service'. The militaristic connections are often subtle in this voluntary youth programme, but nevertheless it encourages duty-bound performances of active citizenship by the nation's youth in preparation for 'full' citizenship status. This argument is more fully developed in Chapter 7 through its discussion of the global geopolitics of 'compulsory volunteering'. Finally, this chapter's examination of the growing role of NCS within schools supports the book's third core argument on the blurred boundaries of formal and informal education. Despite being pitched as an alternative to school, NCS relies on these spaces for recruitment and has become enrolled into the wider state apparatus of character education through pathways in formal education. The encouragement for school leaders to embed youth social action in their curriculum, ethos and extracurricular activities further demonstrates these shifting geographies of education.

NCS continues to grow within England and Northern Ireland and remains a voluntary scheme for teenagers at the time of writing. However, the trends it represents around youth volunteering and civic virtues have global antecedents and international synergies that are examined in the next chapter. Chapter 7 traces current and future trends in the global character education movement and outlines how debates in this area beyond the United Kingdom have a growing (geo)political importance. The discussion that follows brings together examples from both formal and informal learning spaces to map the global geographies and geopolitics of character, citizenship and values.

Notes

1. DCMS (2019) Twitter Post @DCMS, 30 July 2019. Available from: https://twitter.com/DCMS/status/1156212257665236993.
2. Animated Whiteboard Video Diary – NCS Week 1 – Summer 2015, ES/L009315/I.
3. Interview with NCS Graduate, February 2016, ES/L009315/I.
4. Interview with NCS Graduate, February 2016, ES/L009315/I.
5. Interview with NCS Architect, March 2015, ES/L009315/I.
6. Interview with NCS Architect, March 2015, ES/L009315/I.
7. Interview with NCS Architect, March 2015, ES/L009315/I.
8. The 'University College and Admissions Service' supports applications to Higher Education Institutions in the United Kingdom.

7 Character, citizenship and values

From national debates to global geopolitics

Introduction

This chapter turns the book's focus towards the contemporary geographies of character education in a range of different national contexts. The recent (re)turn to values and virtues in England over the last decade has been inspired by several global examples, outlined in Chapter 3. This chapter builds on those multi-scalar geographies but goes beyond tracing global inspirations to instead map current developments in this area across the globe and their geopolitical dimensions. A global perspective on these growing trends helps to elucidate the wider politics of character, citizenship and values. This landscape of character education across the world is an uneven terrain, with most developments in this field still largely located within the United States and Asia. However, this chapter does engage with a range of examples in these and other international contexts across a wide political spectrum. The chapter's critical analysis is divided into four interlinked sections that contribute to the book's wider arguments on the geographies and geopolitics of character education and the shifting boundaries of formal and informal learning spaces and practices.

First, this chapter examines the wider global geographies of character education, outlining key recent trends in different national contexts. These include developments in countries with more established track-records of character education, such as the United States and Singapore, as well as newer revivals in countries such as Taiwan and the United Arab Emirates (UAE). Through a critical examination of current trends around curriculums, assessments and the built environment, the chapter demonstrates how moves in the English education system and wider civil society discussed in this book are not isolated or unique, but rather part of a much broader global character education movement and attempts to govern through pedagogy.

Second, this chapter outlines and explains the growing role of international networks in the character education movement. These networks are focused on advocacy, research and mutually beneficial partnerships that shape the moral geographies of character education and reinforce the

DOI: 10.4324/9780203733066-7

growing trend of positive psychology within educational spaces. This section also demonstrates unequal access to such international experiences and exchanges for children and young people, and by extension inequalities in the accrual of cultural capital.

The third section of this chapter critically examines the global geopolitics of youth volunteering and character-building programmes for children and young people. The book has thus far discussed some of these connections, for example the presence of military ethos providers (Chapter 3) and encouragement of Army Cadet units (Chapter 4) in English schools, as well as historical character-based visions for uniformed youth (Chapter 5) and the establishment of a 'modern' voluntary national service with NCS (Chapter 6). This section pushes these ideas further to explore the growing militarisation of 'compulsory volunteering' within different national contexts, especially as part of recent election campaign pledges in France and the United States.

Finally, this chapter discusses the wider governance of 'good character' through several examples across the world. This section demonstrates the political relevance of taking character education seriously, given the extension of several of its ideas on behaviour, ethics and morals beyond formal and informal learning spaces and into the home and public space. The chapter ends by reflecting on these dynamics for all individual citizen-subjects, not just children and young people.

Global geographies of character education: current trends and future directions

Spohrer and Bailey observe how there has been a "recent resurgence" (2020: 562) of character education and values-based rhetoric in countries such as Australia and Canada, as well as continued growth in the United States and Singapore which boast long-standing traditions of moral education. This section maps these global geographies of character education, both within these countries and beyond, and highlights three significant current trends and developments that support the book's argumentation. Furthermore, the section hints at some of the potential future directions of travel for the character agenda in England.

The first analytical observation is the increased targeting of character education initiatives towards younger children and the firmer enrolment of parents into these activities. For example, the Ministry of Education in Singapore reviewed its character and citizenship education (CCE) syllabus in 2019, ahead of planned changes in 2021, with the intention of a stronger focus on moral education for children at a lower primary level. Singapore introduced CCE in 2014 as the latest chapter of its historical relationship with character or 'values education', reflecting state ideologies of multiracialism and communitarianism (Tan and Tan, 2014). The existing CCE curriculum has a primary, secondary and pre-university syllabus, but the potential changes to a more values and moral-based curriculum at primary level with younger children are noteworthy. Indeed, early signs of the targeting of younger age groups within England are evident through the Department for Education (DfE) 'My Activity Passport' of nature-based encounters for primary schools to foster gritty resilience in pupils (Chapter 4) and social action programmes for younger children to encourage generosity and neighbourliness (Chapter 6).

The recent review of CCE in Singapore seeks to emphasise "core values like respect, responsibility, resilience, integrity, care and harmony", described by Education Minister Ong Ye Kung as the foundation of a "kind society" (cited in Kurohi, 2019). This framework views these characteristics as shared national values but also shared family values, reflecting wider ideas in Singapore about citizenship, national identity and familial belonging (Tan and Tan, 2014). Mr Ong recently emphasised that these values promote educational pathways and careers for young children, but also "guide the relationships that they have with their families, peers, teachers, and the community that they live in" (cited in Kurohi, 2019). This policy therefore also invokes the notion of normative family life for good citizenship. Ministry of Education guidelines about primary school education in Singapore explain there is a 'Family Time' segment in CCE as well as other periods in school that "promotes parent child bonding through suggested activities" (MOE, 2020: 17). For example, a recent guide for parents includes a QR code of "Tips on how you can support your child's educational journey" (MOE, 2020). Whilst these trends of parenting advice, classes and education are identifiable in England and other contexts in the Global North (see Holloway and Pimlott-Wilson, 2014b), this example from Singapore demonstrates recent moves to extend the reach of the state into home space and family life through specific pathways of character education.

The second recent development to emphasise in this section on the global scale is the increased measuring and monitoring of student's abilities in, or aptitude for, character education. This idea was discussed in Chapter 4 in relation to attempts by US educational psychology practitioners to measure and quantify an individual student's grit, pluck and resilience. More broadly, there is growing evidence to suggest that in countries where it is popular, data-driven approaches to character education assessment will become more prevalent. For example, the United Arab Emirates (UAE) introduced moral education as a compulsory subject across all schools in 2017 and has recently moved towards standardised assessment. The four pillars of its moral education curriculum aim to "build character, instill moral outlook, foster community and endear culture" (Gulf News, 2019) as part of a wider framework that increasingly promotes mindfulness and social-emotional learning in individual schools (Masudi and Zaman, 2019). Strikingly, UAE have developed a computer-based Moral Education Standardised Assessment (MESA). In 2019, over 10,000 students in 78 schools were measured on their "understanding of moral education, the effectiveness of the curriculum, and its impact on the development of students' personalities" (Gulf News, 2019) making this example one of the most formalised assessment mechanisms for character education in the world. More broadly, the character agenda is increasingly part of wider debates on the datafication of education (Finn, 2016; Williamson, 2020) and in particular the enumeration of non-cognitive skills of students. Williamson (2021) argues that this quest, fuelled by transnational networks and interest groups, creates 'psychodata' that makes "'personality' an international focus for policy intervention" (2021: 129). Overall, ethical debates about analytics technologies in education and the wider 'neurogovernance' of 'brain data' (Pykett, 2015; Williamson, 2019) look set to continue to shape the wider cultural politics of (character) education.

More broadly, the educational ambitions and performance of individual nation-states are increasingly understood through data-driven technological advancements. For example, the current education and culture minister in Indonesia recently announced his technology-based visions to improve overall educational standards, which notably included a feature on 'competency and character' (Amirrachman, 2019). Although schools in England and the United Kingdom do not currently assess individual children and young people's knowledge or aptitude in the specific area of character, Chapter 4 hinted at some early moves towards the evaluation of individual school's character, for example through the recent new Ofsted framework (see also Winton, 2008 on this debate in Canada). In addition, it is worth emphasising here that in the last decade, character education has been a route to wider educational technology developments in various national contexts. For example, the growth of plasma TVs for instructional videos as part of Ethiopia's 'Civic and Ethical Education' curriculum in secondary schools (Semela, Bohl and Kleinknecht, 2013). Although the model of Satellite Educational TV broadcasted lessons has been patchy in effectiveness, this example demonstrates how wider educational changes, technological developments and digital inequalities can be read through the lens of character education.

The third analytical observation to highlight in this section is how character education is increasingly marked in the built environment across diverse national contexts. In Taiwan, for example, more than 50 schools recently received plaques from the National Parents' Association, reading "li yi lian chi" 禮義廉恥 translated as "propriety, justice, integrity and honor" (Mengchuan and Hsiao, 2019). These words were common features in Taiwanese schools decades ago, and these recent gifts therefore represent a nostalgic (re)turn to these ideas, cemented in the physical landscape of school buildings. Indeed, despite educational reform and social change in Taiwan (Lee, 2007), there has also been a renewed focus on character, values and moral education.

The role of the built environment in these debates may seem banal and mundane, but these examples often reflect political insecurities connected to the character education agenda. In Australia for example, a highly symbolic 'Flagpoles for Schools Programme' was introduced in 2005 that provided schools with funding to install and display the Australian national flag. Cranston et al. argue that this scheme should be read as an attempt by the Federal government to "impose a particular set of values on Australian schools" (2010: 188). 'Flagpoles for Schools' was part of a wider drive that same year to introduce a 'National Framework for Values Education in Australian Schools' (Jones, 2009). This apparently responded to parental and public fears that the previous curriculum was too 'values-neutral', or as acting Minister for Education Peter McGauran described, was "hostile or apathetic to Australian heritage and values" (cited in Clark, 2007). Instead, a new values framework was embedded with an overall mission of "care and compassion, doing your best, fair go, freedom, honesty and trustworthiness, integrity, respect, responsibility and understanding, tolerance and inclusion" (Cranston et al., 2010: 189). There are clear synergies here with ideas of school ethos and mission statements in English schools discussed earlier in this book, but notably 'fair go' in this example is an attempt to capture a colloquial and seemingly nationally specific Australian value. This identification and promotion of apparently unique national values is a popular current trend in character education circles, to emphasise national specificity alongside universal virtues. However, the extent to which there is something discernably 'Australian', or indeed 'Taiwanese' or 'British', that children and young people learn from curriculums, flagpoles or plaques, is questionable. This debate sits at the heart of critiques about the policy and practice of FBV. These values attributed to Britain and British citizens of "democracy, rule of law, individual liberty and mutual respect and tolerance" (DfE, 2014d: 5) have been politicised at the same time as character education has flourished. This example therefore reflects wider debates on the contested notion of a collective 'national' character (Inkeles, 1996; Mandler, 2007), which are increasingly expressed in the built environment.

Overall, this section has outlined three central trends in character education in schools that cut across different national examples: first, targeting younger children through curriculum initiatives that also enrol parents and families; second, the increased use of testing, measuring and monitoring character through digital technologies; and finally, cementing character-based national values within the built school environment. Whilst elements of these three current developments can be seen in arguably the most popular global example of character education infrastructure – the charter school movement in the United States – there is a significant final reflection in this section on that national context. There are encouraging, if not wholly popular, recent moves within the US charter school system to acknowledge structural barriers and some of its problematic language around grit, discussed in Chapter 4, in the context of racial inequalities. The Knowledge is Power Program (KIPP) School Movement announced in 2020 that it would be removing its original slogan of 'Work hard. Be nice' as this "places value on being compliant and submissive" and "supports the illusion of meritocracy" (cited in Mathews, 2020). This recent policy change has been the focus of a polemic 'culture wars' style debate. Indeed, the editorial board of the *Wall Street Journal* described KIPP's decision as 'woke nonsense' and points to its existing success in teaching Black and Latino students. However, Love (2019a) has outlined the structural inequalities and racism within this landscape and wider educational injustice in the United States. The example of this slogan change for a major character education organisation highlights how in countries with established track records in this area, their policies and practices are not static but part of wider, shifting moral geographies of education. The next section examines a growing number of global networks and advocacy groups centred on character education that further demonstrate the politics and inequalities of this wider movement.

Global networks and advocacy for character education

Many geographers and social scientists have shown that educational ideas, policies and activities – as with any form of knowledge production – have global mobilities (i.e. Jöns, Meusburger and Heffernan, 2017; Brooks and Waters, 2018). At the most basic level, there can be knowledge exchange about character education between individual nation-states, for example Scotland's decision to pilot Australia's 'Bounce Back!' programme (Axford, Blyth and Schepens, 2010). More broadly however, there are a range of international networks and interest groups that promote, advocate and lobby for character education, acting as global hubs that distribute resources and research to further grow the movement. These global networks often adopt one of two models: first, an advocacy model that champions character education's benefits and evidence base to new audiences; and second, a business-style model that facilitates international partnerships and exchanges between schools with a like-minded character ethos.

An example of the first type of network is the International Positive Education Network (IPEN) founded in 2014. Their team and advisory board include practitioners, business leaders and senior researchers. IPEN defines its purpose as "a global network...to promote positive education and bring about a paradigm shift in perceptions of the purpose of education" (IPEN, 2020). Notably, the network uses positive education rather than character education specifically, reflecting its wider educational philosophy and focus on mindfulness. Positive psychology is a growing feature of contemporary neoliberal social life, especially in policymaking and workplace training, for example that focuses on optimism and happiness (Pykett and Enright, 2016; Whitehead et al., 2016). This academic discipline and industry, along with many of its key champions, are what Jerome and Kisby (2019) have identified as a wider 'policy community' of character education (see also Allen and Bull, 2018). The aim in this section is not to repeat the detailed configurations of these networked relationships, but to acknowledge these exist and connect supporters, for example through interactive maps, international conferences and a growing 'industry' explored shortly.

The second type of global network offers a paid service to schools in return for the benefits of partnerships and exchanges that enhance the character of its respective pupils. This business model echoes some of the wider politics of school choice, privilege and cultural capital discussed in Chapter 4, but here operating at the international scale. For example, the organisation 'Round Square' brings together "likeminded schools on an international stage" to facilitate effective character education. This includes supporting project trips and international adventures as well as digital engagements and values-based school activities centred on a philosophy of IDEALS, which stands for "Internationalism; Democracy; Environmentalism; Adventure; Leadership; Service" (Round Square, 2020). They promote values-based and experiential learning via partnerships with 200 schools worldwide, including those in India, Korea, Brazil, Canada, Australia, Argentina, Kenya, Pakistan, New Zealand and the United States. Schools first apply to be a candidate school of Round Square, which describes itself as a community and non-profit charity, with a further process of application for global membership after two years or when 'ready to progress". At that stage, a membership fee is payable to access Round Square benefits. The types of membership benefits and exclusivity these and other similar networks offer to individual schools further demonstrate the multi-scalar geographies of character education.

More broadly, these two types of network are part of a wider character education 'industry". Indeed, there are synergies here to other global industries and knowledge economies, such as the migration industries (Cranston, Schapendonk and Spaan, 2018). Studies in this area go beyond mapping organisational infrastructures to understand how "processes of migration become an economy; the production and circulation of knowledge, the offering of services and so on" (Cranston, et al., 2018: 544). The networks outlined in this section commodify expertise on character education through workshops and conferences, reflecting how knowledge economies are performed and their geographies (i.e. Cranston, 2014 on trade shows). The character education industry also operates within wider global flows of philanthropic support. For example, Allen and Bull (2018) have traced how funding from the US-based neoconservative John Templeton Foundation has supported some of the research of the Jubilee Centre in the United Kingdom (see also Jerome and Kesby, 2019). The scale and reach of the Templeton Foundation is much broader, for example the Templeton World Charity Foundation has funded research and evaluated emerging character education projects across the globe (i.e. de Soria et al., 2018 on Mexico, Columbia and Argentina). Overall, these interconnected global networks of research, advocacy and philanthropy further demonstrate the wider politics and moral geographies of the character education movement.

Compulsory volunteering? The global geopolitics of national service

This section examines wider global trends on youth volunteering, including their growing geopolitical dimensions, building on the focus of Chapter 6 on National Citizen Service (NCS). Youth volunteering is often held up as a universal good, a panacea to a wealth of social, economic and political issues and the antidote for social cohesion and apathetic young people. In recent years, there have been an increasing number of election pledges around the world focused on youth volunteering and 'national service', intended to foster civic virtues of duty and the promotion of national values. These have either been commitments from politicians to (re)introduce military formats of conscription for teenagers or to create more civic-minded programmes of service learning. The boundaries between these two aims are often blurred, as the following examples in this section demonstrate.

In May 2019, Emmanuel Macron launched a new national civic service pilot for French teenagers – Service National Universal (SNU). His original vision of military-based national service was part of an election pledge, but the reality has been a programme focused on the civic virtues of youth volunteering but that will eventually become compulsory. Macron's scheme has already faced criticism for its overtly nationalistic performances of recruits singing the national anthem and its disciplinarian tone of banning mobile phones (Chrisafis, 2019). The eventual plans for this programme in France are ultimately rooted in the idea of *compulsory volunteering*. However, this is a complete oxymoron as there is nothing voluntary about forced volunteering. Indeed, this example links to recent debates in the United Kingdom about the possible exploitation of young volunteers in state-led welfare systems, where benefit claimants for jobseekers' allowance have been tasked with unpaid work placements in order to retain this financial support (Ainsworth, 2012). In Yang's (2017) study of 'compulsory volunteering' in Canada via a mandatory high-school community service programme, they found that whilst participation rates clearly went up, this was not sustained after the programme and there was no evidence for increased altruism. The SNU in France will soon become compulsory, revealing national insecurities about 'returning' youth to seemingly lost ideals.

A lively social media debate on this theme of compulsory volunteering emerged in the United Kingdom in Spring 2019 during the Conservative Party leadership contest. Rory Stewart MP's campaign and rapid rise to fame included a pledge for a compulsory NCS. Stewart planned to '–launch' a scheme for 16-year-olds across all parts of the United Kingdom. However, he bizarrely failed to mention that NCS had existed for nearly a decade under his own party's leadership. Nevertheless, Stewart's vision was for a universal and compulsory scheme, which did mark a shift change and one more akin to Macron's vision in France. Stewart spoke of his own military background as part of the rationale but did stress his idea was a civilian version intended as a 'bonding' experience for teenagers. Stewart's idea and campaign was short-lived, but again demonstrates in the context of this section how youth volunteering can become entangled with wider (geo)political debates on character, citizenship and (national) values.

The most well-known example of this debate on 'modern' national service in recent years is in the United States, where several democratic candidate hopefuls for the 2020 Presidential campaign pitched ideas around service learning. For example, senator Elizabeth Warren outlined a plan for a modern 'civilian conservation corps' for America's National Parks, combining service learning with environmental themes (Prior, 2019). Furthermore, early democratic candidate Pete Buttigieg made 'a new call for service' (Myong, 2019) as part of his campaign, intended to foster an expectation of volunteerism among young people. The former mayor of South Bend, Indiana's voluntary plan was noteworthy in relation to the politics of youth volunteering. It would have extended the existing scheme of AmeriCorps to reach 1 million American teenagers engaging in service by 2026 and was framed as helping to repair the 'national social fabric' (Rodriguez, 2019). Buttigieg's vision was also reflective of youth-focused activism, with his plan for new 'Corps' including 'Climate Corps' and another scheme-based around communities and mental health. His plan for an optional year of paid service before work or college was also significant for including incentives for voluntary participation, including time served resulting in student debt relief (Ismay, 2019). This further oxymoron of *paid volunteering* within this debate also raises interesting questions about the wider geographies and geopolitics of youth volunteering.

Overall, there is no doubt that national service style schemes are currently having a renaissance in public policy and election pledges, yet it is important to take a step back in any analysis. For example, in the United Kingdom, there have been periodic calls for the 'return' of national service since the end of military conscription in the early 1960s. This nostalgic debate, often framed as an intergenerational tussle, is often rehearsed during times of national crisis, usually with reference to youth crime, social cohesion or low levels of political and democratic participation. The timing of these cyclical debates on national service also tends to reflect the current political mood, often as an adult response to what young people are seen to 'need' and what the nation feels it needs from its future citizenry. It is also important to remember that many countries around the world still have compulsory military conscription. There are some well-known examples of conscription such as North Korea and China, but countries as diverse as Norway, Greece, Singapore, Austria and Finland have age-based and in most cases gender-based forms of military national service. The latest champions of new or rebranded voluntary service programmes such as Buttigieg in the United States and Stewart in the United Kingdom are at pains to stress that their schemes are non-military. However, there will always be these echoes of militarism in any new scheme for as long as their programme names continue to include the words 'national service' or talk of 'corps', 'enrolling' and 'recruits'.

Debates on national service and the politics of youth volunteering across the globe reveal the deeper hopes and fears of individual nation-states and their respective visions for the character of their (future) citizens. The next section of this chapter examines how these and other powerful ideas about character, citizenship and values 'spill over' from educational spaces into wider state governance of all citizens, not just children and young people. It draws on a range of examples in different national contexts to demonstrate the wider contours of moral geographies around behaviour, conduct, integrity and (national) etiquette, elucidating the wider political relevance of this debate.

The governance of 'good character': individuals, families and states

Across the globe, character education can be a key political tool in shaping and governing citizens. However, the relationship between character, citizenship and values is not only situated within educational sites and settings but can filter from these spaces into wider state governance in homes, families and public space. Furthermore, several examples in this section outline how the governance of 'good character' can extend to all citizen-subjects regardless of age. This critical observation is not limited to criminal or legal contexts; indeed, the introduction to this book referred to 'Good Character' tests by the UK Home Office as an example of the place of character within wider legal state apparatus. Rather, this section is concerned with a more diffuse, ethical and behavioural emphasis regulating individual citizen-subjects which in some cases operate in tandem with the character education agenda.

Russia introduced character education in September 2020 with a focus on 'social, cultural, spiritual and moral values', which on the surface echoes similar sentiments to the (re)turn to character in England. However, this new legal status for character education in Russia also includes legal amendments that children should learn:

> the rules and norms of behavior adopted in Russian society in the interests of the individual, family, society and the state. It should cultivate a sense of patriotism and civic consciousness; cherish the memory of the nation's defenders, instill respect for law and order, workers and the older generation.
>
> (TASS, 2020)

The introduction of these new measures by President Vladimir Putin has a clear focus on building character, but it also explicitly focuses on 'upbringing' and extending the state's reach into home spaces.

This focus on children and young people's behaviour through the lens of character also extends to public spaces. For example, in the British Virgin Islands (BVI), there is a legal student code of conduct and rules-based set of guiding principles on character. Introduced in 2006, the Education (Student Code of Conduct) Rules were recently reviewed after a perceived slippage in young people's behaviour (BVI News, 2019). This example illustrates how a set of rules for students can have a wider influence beyond school walls and permeate other spaces, for example the BVI's review sought to address "unacceptable behaviour" such as "public acts of indiscipline that have consistently made rounds on local social media platforms" (BVI News, 2019). This focus on public space also emerges within the recent push for citizenship education in Zimbabwe. This utilised a number of similar justifications based on the public's perceptions of young people's behaviour, which was ascribed to "lack of citizenship values, relevant ethics, morals and individual and collective responsibility towards property" (Sigauke, 2012: 215). The Presidential Commission on the curriculum drew on long-standing social constructions of young people and general assumptions, framing education as a potential 'tool' for character formation. However, as Sigauke's (2012) study highlights, students in Zimbabwe prior to the introduction of any curriculum reform were indeed knowledgeable on citizenship issues, but their civic engagement was curtailed by issues of trust and wider political tensions. Furthermore, schooling more generally is seen by nation-states as having enduring power to combat moral ills and avoid moral 'pitfalls'. This argument is made by Dungey and Ansell (2020) in their recent study of education in Lesotho in Southern Africa. The authors stress that as well as navigating economic and physical uncertainties, school still plays an important role in preserving and protecting moral standards, perceived as helping to avoid risks such as loitering, alcohol, drug-taking and extra-marital sex. Their study clearly demonstrates the moral geographies of education and how the governance of 'good character' extends its reach into everyday life.

Adult citizens in some national contexts are also included in wider character education 'projects' that extend beyond the classroom or campsite and into wider citizenry. For example, China introduced new morality guidelines for all citizens in 2019. The 'Outline for the Implementation of the Moral Construction of Citizens in the New Era' includes a focus on civic education and 'national etiquette' (Kuo, 2019) as well as good behaviour. These guidelines of course reflect the ruling Chinese communist party but are a stark example of how citizenship can be constructed along moral boundaries and how ideas of etiquette and manners are recast as everyday embodied performances of national identity for children and adults.

Another example of character and values-based ideas in education 'spilling over' to adult citizens, and parents in particular, is the recent college scandal in the United States. A number of criminal trials took place in 2019 in relation to fraudulent college admissions applications. The high-profile case of US actress Felicity Huffman, who pleaded guilty to fraud and conspiracy, shone a media spotlight on this hidden but wider systemic issue. Huffman secured entry to an elite school for her daughter through hiring an admissions consultant to arrange for exam SAT scores to be altered and falsified through bribery (BBC, 2019). The actress was eventually sentenced to two weeks in prison, a $30,000 fine and 250 hours of community service, referencing her "desperation to be a good mother" amid an increasingly competitive admissions landscape (BBC, 2019). The wake of the US college scandal put the character and integrity of individual parents firmly under the spotlight, with 51 individuals charged and fined in this specific incident, many of whom also bribed coaches to falsify sport-related achievements and where children were often unaware of attempts to doctor SAT results (BBC, 2019). The lesser custodial sentences received by the mostly white elite wealthy parents in this fraud scandal sit in stark contrast to the over-zealous criminalisation of poor, often Black and Latino parents in the United States for education-related crimes such as truancy or 'stealing education', where children are registered for schools in a different location to a family's residential address (Goldstein, 2015; Baldwin Clark, 2019). Overall, the potential (il)legality of parents' actions in securing educational access and prestige reflects complex constructions of the 'good student' and 'good mother' as well as demonstrating the wider relationship between white privilege, class and education in the United States (Warikoo, 2016). More broadly, these examples and their moral geographies illustrate the salience of character for understanding the cultural politics of education and the wider governance of 'good character'.

Conclusion

The global map of character education activity is uneven and not universal across the world. For example, citizenship education is far more popular in some nation-states, as outlined in Chapter 2. Where an educational focus on morals, virtues and values is present, these trends towards character are growing, with recent developments in a range of national contexts worthy of further scholarly attention. This chapter has critically examined the global geographies of character, citizenship and values through four distinct yet inter-related sections that contribute to the book's overarching arguments.

First, the chapter provided a critical synthesis of recent developments within the character education movement across different national contexts. These examples of governing through pedagogy included the extension of moral-based learning towards lower primary-school age groups, moves toward the datafication and assessment of character education, as well as national values and character traits inscribed into built school environments. Second, this chapter outlined the influential role of international networks and interest groups for promoting character education and shaping a growing industry. These activities strengthen the wider positive psychology movement as well as entrenching inequalities via exclusive global school partnerships and exchanges. The first two sections of this chapter therefore contribute to the book's central argument that geography matters in understanding character education and the wider relationships between character, citizenship and values. The sections have demonstrated the multi-scalar geographies of education, for example through global networks, as well as how recent trends in character education are expressed in subtly different ways in different nation-states. Whether new curriculums in Singapore, new flagpoles in Australia or new plaques in Taiwan, or new assessments and teaching technologies in UAE or Ethiopia, place matters to the geographies of character education. These dynamics are also shaped by socio-spatial relations, for example character education's place within families, the built environment, or in addressing or perpetuating wider social and economic inequalities.

Third, this chapter examined the global trends towards 'modern' national service programmes for young people and their enduring place in election campaigns and state policies. This chapter therefore contributes to the book's central argument on the growing geopolitical dimensions of character education. It has elucidated how these schemes with militaristic overtones, or indeed formal conscription, are framed in terms of promoting 'national values' and repairing the 'national social fabric'. The chapter also raised some provocative questions around the notion of 'compulsory volunteering' and how these tensions have wider geopolitical importance.

Finally, this chapter exposed how the governance of good character can 'spill over' from sites of formal and informal education into wider spaces and audiences. A series of instructive examples from different national contexts demonstrated the politics of behaviour, etiquette, integrity and honesty within wider character projects aimed at children, young people and adults. This discussion therefore contributes to the book's final core argument on the blurred boundaries of formal and informal education, utilising examples from both school spaces and youth volunteering programmes, but also highlighting the wider reach of these ideas beyond educational sites into the home, public space and adult citizenry.

Overall, this chapter has extended the book's central focus on England to map key trends and current developments related to character education across the globe. England and the United Kingdom are of course part of these wider multi-scalar geographies and are not isolated or unique from these specific debates. However, the global perspective in the chapter has further demonstrated the politics of character, citizenship and (national) values both within and beyond educational settings. The next chapter offers a series of wider conclusions and final reflections on this book's arguments and contributions. Chapter 8 draws this book and its mapping of the moral geographies of education to a close.

8 Conclusion

Character education has 'bounce back'. Over the last decade in England, through successive governments and cabinet reshuffles, it has demonstrated its own 'stickability' and gritty determination to persevere. This book has explored the moral geographies behind this renewed character agenda and its complex genealogy, from reimagined school ethos to the rise in extracurricular activities, and from established voluntary youth movements to new state-led programmes encouraging social action. The research underpinning this book and shaping its central arguments has contributed to academic debates and literature introduced in Chapter 1, especially on the geographies of youth citizenship, geographies of education and geographies of children, youth and families. First, the book argues that geography matters in understanding the character agenda, both in terms of the moral geographies that spatialise ideas about behaviour and youth citizenship, as well as the multi-scalar geographies of its practices. Second, the book argues there is a growing geopolitical dimension to character education that is vital to consider in any analysis, as children and young people's experiences of character education are increasingly shaped by military ethos and spaces. Finally, the book argues that the boundaries of formal and informal education have become increasingly blurred through the character agenda, shaping and reflecting changes in the wider relationship between the state and civil society. Overall, the book's analysis of historical and contemporary material and the data set outlined in Chapter 2 has demonstrated these arguments as well as the wider significance of character, citizenship, and values.

This chapter provides a detailed summary of the book's three overarching arguments in the context of its wider aims and objectives. The chapter highlights cross-cutting themes and emphasises its contributions to knowledge. The chapter ends with a series of reflections on the place of character in social and political life during the current context of COVID-19.

The first central argument of the book that geography matters in this politically vital debate has been demonstrated in two central ways. First, by foregrounding the *moral* geographies of character education in every chapter and using this entry point to trace this wider landscape. This book's examination of character education activities in schools and youth

DOI: 10.4324/9780203733066-8

volunteering programmes, whilst seemingly mundane or perhaps even niche, actually reveals the contours of a much broader moral geography of the relationship between states and their citizens. The book has illustrated the role of moral values, virtues and behaviours in defining the norms and contours of good citizenship over time and space: the geographies of youth citizenship. Characterful attributes and at times ambiguous traits such as grit, gumption, generosity and neighbourliness were the focus of the book's detailed critical analysis of how these and related ideas are spatialised as they 'take shape' in the classroom or campsite (Chapters 4–6). Second, the book illustrated how and why geography matters by excavating the *multi-scalar* geographies that shape and help accomplish character education in practice, particularly in Chapters 3 and 7. The book outlined the global movement and international networks of character education, inspiring national ambitions and co-opted into declarations of national values. In the United Kingdom, these are re-shaped through the devolved governance of education, and in England, have been translated into policies intended to foster a 'character nation'. A (re)turn to these ideas under the 'umbrella' of character is then realised in practice at the local scale, shaped by the uneven and increasingly autonomous dynamics of educational restructuring in England, and emerging in the built environment and spaces of local schools through the three pathways of curriculum, ethos and extracurricular activities. These ideas are eventually performed at the scale of the embodied actions of children and young people, whether in schools (Chapter 4), uniformed youth movements (Chapter 5) or National Citizen Service (NCS) (Chapter 6). Overall, a geographical approach to character education attentive to the role of scale and how place matters in shaping moral ideas has injected a much-needed geographical focus into interdisciplinary debates on character education.

The second key argument of this book is that the geopolitics of the character agenda are an increasingly important dimension. The discussions on national values, (in)securities, military ethos and different examples of uniformed youth across this book reveal the significance of this debate and the wider geopolitics of governing through pedagogy. Character education is not just a pet project of the current UK Government, but rather a continual project by the state to shape and govern the behaviour and conduct of (young) citizens and future adult citizenry. The book has illustrated this argument through the growing place of military ethos providers (Chapter 3) and Army Cadet units (Chapter 4) in English schools. These geopolitical dimensions have also been illuminated through the examples of voluntary spaces of informal education, for example the imperial visions of duty-bound uniformed youth, ideas reworked in the latest call from The Scouts to 'be resilient' as well as 'be prepared' (Chapter 5) and the advent of a modern 'national service' established by UK Government in the last decade with a focus on civic virtues and social action supporting the transition to adulthood (Chapter 6). The book has also, particularly in Chapter 7, outlined the global dynamics of these trends for children and young people and how these ideas can 'spill over' into the wider state governance of behaviour and good character. This book has also outlined some clear synergies between the past and present in its analysis of the re-imagining and re-envisioning of a 'character nation'. It argued that the UK Government clings to idea(l)s from the past as part of its future vision for education. Grit and gumption have been recycled, pluck and zeal emerge once again, and such historical ideas become positioned alongside the new buzzwords of resilience, stickability and bounce-back. In this wider agenda, the allure of uniformed youth and military ethos are back in fashion again, not just in England but as Chapter 7 demonstrated, a range of international contexts.

Finally, the third central argument of this book is that the boundaries of formal and informal education have been reshaped through the renewed character agenda over the last decade. This furthers understandings of the contemporary processes and spatialities of education more broadly. The book argues that the (re)turn to character in England in the last decade can only be fully understood through considering both learning environments. The chapters have illustrated the blurred boundaries and slippage between formal and informal education spaces with a moral compass, for example the Department for Education (DfE) character education pilots through a range of charities discussed in Chapter 3, the focus on extracurricular activities in schools in Chapter 4, the growing relationship between The Scouts and DfE outlined in Chapter 5 and the increasing links between NCS and schools through social action discussed in Chapter 6. The study of the relationships between these spaces and philosophies, rather than considering them in isolation, contributes to wider understandings on the geographies of education and interdisciplinary debates. For example, the book has traced how these institutional geographies shape wider landscapes of childhood and youth, as well as children and young people's lived experiences and inequalities in this shifting terrain of compulsory and voluntary activities. Furthermore, the book has used this lens to demonstrate the wider changes in the relationship between the state and civil society in the last decade, especially in austere times.

Overall, the three core arguments of this book have elucidated the relationships between character, citizenship and values. The intention of the book was never to advocate for a particular approach to character education, or to dismiss one philosophical tradition in this field and embrace another, but rather to map its emergence, spatialities and moral geographies. Likewise, this book has never tried to argue that kindness or being a 'good neighbour' should not be taught in schools or youth organisations. However, this book has asked some more provocative questions about character education that challenge the current dominant narrative that it is a universal good for young people, and increasingly younger children. The discussion in Chapter 4 in particular asked whether the moral norms that character education currently proposes around grit and gumption could create a further entrenchment of the idea that young people are 'not good enough', a defining feature of neo-liberal times. Indeed, this book ends with a call for closer attention to how the 'new' shift towards moral education could be demoralising, and with a call for future research on its current developments in different national contexts, many of which were synthesised in Chapter 7. There are constant political debates around the world about how to improve the educational and moral standards of children and young people. The debates around character education in England are therefore connected to bigger questions about the purpose of education itself, whether its aim is training the future workforce, a driver for social mobility, personal and social development, or to instil nationally specific values and shape future citizenry. Character education is also currently competing with pushes from a wide range of campaigners, educators and activists for the UK Government to pursue a stronger focus on political literacy, democratic education, anti-racist education, social justice, or to re-invigorate citizenship education. These wider cultural values and cultural politics of education are increasingly heightened at the time of writing with the unfolding COVID-19 crisis.

COVID-19 is radically reshaping the boundaries of education at the local, national and global scale. This impact is most obvious through the rapid rise and inequalities associated with online home-schooling, but also the changing practices of youth organisations such as The Scouts to promote 'The Great Indoors'. I do not want to end this book with a premature assessment of the current global pandemic's impact; indeed, it is far too early to tell the long-term effects on children, young people and families. However, there are some striking examples within the United Kingdom during this unprecedented time that demonstrate the significance of character and its continued place within political and social life. Two illustrations stand out in this context for highlighting the wider 'work' character does and its relevance to current and future academic debates.

First, during this recent period, there has been a call for character to become a national mindset or spirit needed to see the country through the COVID-19 crisis. For young people, these ideas have been encapsulated in calls from NCS to "Keep Doing Good" during this period as "our way of banding together to help get the country back to business". This rallying call for social action and volunteering has been pitched to young people as "helping to rebuild your local community" but also that "you'll pick up some new skills and get all of our futures off to a good start" (NCS Trust, 2020). More broadly, all individual citizens have either been framed as possessing or lacking a gritty determination and resolve to 'weather this storm'. This governance through grit – whereby the state articulates that the nation itself needs to be resilient and be prepared – draws on many of the ideas examined in this book. There is a certain irony however in observing political leaders calling for such character traits and virtuous moral behaviour, given Prime Minister Boris Johnson's recent unlawful proroguing of parliament and his defence of special advisor Dominic Cummings after lockdown breaches.

More broadly, the 2020 Presidential Election in the United States between Joseph Biden and Donald J. Trump can also be read through the lens of character in a time of COVID-19. In England, whilst the DfE's current character education plans and any future developments are clearly paused during the immediate COVID-19 crisis, the place of character, citizenship and values more broadly in this political context is striking.

The second example that highlights how moral geographies of character have coalesced in the context of COVID-19 is the UK A-Level results fiasco in Summer 2020. There were continued calls for Education Secretary Gavin Williamson to resign after a so-called 'mutant algorithm', used to determine results after cancelled examinations, caused widespread inequality and impacted the lives of hundreds of thousands of young people (Amoore, 2020; Stewart, 2020). One of the most revealing comments on A-Level results day was those from Conservative peer Lord Bethell. As a cohort of pupils, parents, headteachers and Universities processed the downgrading of centre-assessed grades, Lord Bethell stated on social media that grit and perseverance were more important than A-Level results. He drew on his experience of 'fluffing' formal examinations and instead "learning how to hustle", stating that "Grades are great, but grit and perseverance win every time".[1] The comments were widely critiqued and sparked much debate, given that Lord Bethell attended private school and inherited his seat in the House of Lords (Singh, 2020). Lord Bethell apologised, but the privileged sentiment embodied in this example chimes with the debate at the heart of this book. In particular, the argument in Chapter 4 on the deep-rooted moral geographies of grit and perseverance shaped by class, race and gender. This overall example can be critically viewed through the framing of character, whereby a young person's disappointment of 'mutated' A-Level results should, for some, instead be embraced as a moment to display gritty perseverance and 'bounce-back'; the ultimate character-building experience from the current UK Government. At the time of writing, some journalists and commentators have gone as far as suggesting that "Covid kids can become Generation Grit", emerging "as the most resilient in decades" (Thomson, 2021).

Overall, character remains a popular and powerful idea, even during the current political and public health challenges of COVID-19. This book has demonstrated that there is a long-standing interest in character training and governing through pedagogy both within and beyond formal education and these moral geographies should continue to be mapped. Furthermore, the wider relationship between character, citizenship and values remains a significant debate and lens through which to view wider social and (geo) political debates.

Note

1. Lord Bethell (2020) Twitter Post @JimBethell, 13 August 2020, Available from: https://twitter.com/JimBethell/status/1293905136687804417?ref_src=twsrc%5Etfw.

References

ACE (2020) *Association for Character Education*. Available from: http://character-education.org.uk/.

Ainsworth, D. (2012) 'Compulsory volunteering' for those on benefits, *Third Sector Blog*. Available from: http://thirdsector.thirdsector.co.uk/2012/01/19/compulsory-volunteering-for-those-on-benefits/.

Aitken, S. C. (2001) *Geographies of Young People: The Morally Contested Spaces of Identity*. London: Routledge.

Ajegbo, K. (2007) *Diversity and Citizenship: A Curriculum Review*. London: Department for Education and Skills.

Alexander, K. (2009) Similarity and Difference at Girl Guide Camps in England, Canada and India, in N. R. Block and T. M. Proctor (eds) *Scouting Frontiers: Youth and the Scout Movement's First Century*. Cambridge: Cambridge Scholars Press: 106–120.

Allen, H. (1986) Character Takes on Personality, *The Washington Post*, 5 January 1986. Available from: https://www.washingtonpost.com/archive/opinions/1986/01/05/character-takes-on-personality/d15aeae3-d266-4c8b-a7c2-77cbcbdcb3b6/.

Allen, K. and Bull, A. (2018) Following policy: A network ethnography of the UK character education policy community, *Sociological Research Online* 23 (2): 438–458.

Althof, W. and Berkowitz, M. W. (2006) Moral education and character education: Their relationship and roles in citizenship education, *Journal of Moral Education* 35 (4): 495–518.

Amirrachman, A. (2019) Can technology shape students' character? *The Jakarta Post*, 16 November 2019. Available from: https://www.thejakartapost.com/academia/2019/11/16/can-technology-shape-students-character.html.

Amoore, L. (2020) Why ditch the algorithm is the future of political protest, *The Guardian* 19 August 2020. Available from: https://www.theguardian.com/commentisfree/2020/aug/19/ditch-the-algorithm-generation-students-a-levels-politics.

Anderson, B. (2013) *Us and Them? The Dangerous Politics of Immigration Control*. Oxford: Oxford University Press.

Anderson, B. (2019) Cultural geography II: The force of representations, *Progress in Human Geography* 43 (6): 1120–1132.

Andrews, R. and Mycock, A. (2007) Citizenship education in the UK: Divergence within a multi-national state, *Citizenship Teaching and Learning* 3 (1): 73–88.

Angrist, J. D., Dynarski, S. M., Kane, T. J., Pathak, P. A. and Walters, C. R. (2012) Who benefits from KIPP? *Journal of Policy Analysis and Management* 31 (4): 837–860.

Art of Brilliance (2020) 'Bouncebackability'. Available from: https://www.artofbrilliance.co.uk/training/schools/bouncebackability.

Arthur, J. (2005) The re-emergence of character education in British Education Policy, *British Journal of Educational Studies* 53 (3): 239–254.

Arthur, J. (2020) *The Formation of Character in Education: From Aristotle to the 21st Century*. London: Routledge.

Arthur, J., Kristjánsson, K., Harrison, T., Sanderse, W. and Wright, D. (2017) *Teaching Character and Virtue in Schools*. London: Routledge.

Arthur, J., Kristjánsson, K. and Thoma, S. (2016) *Is Grit the Magic Elixir of Good Character?* Available from: https://virtueinsight.wordpress.com/2016/06/15/is-grit-the-magic-elixir-of-good-character/.

Asthana, A. and Campbell, D. (2017) Theresa May paves way for new generation of grammar schools, 6 March 2017. Available from: https://www.theguardian.com/uk-news/2017/mar/06/theresa-may-paves-way-for-new-generation-of-grammar-schools.

Austin, R. (2020) *Geographies of Youth Leadership: Volunteering and the Young Leaders' Scheme in the UK Scout Association*, PhD Thesis, Loughborough University, UK.

Axford, S., Blyth, K. and Schepens, R. (2010) *Can We Help Children Learn Coping Skills for Life?* Perth and Kinross Council Education and Children's Services.

Baldwin Clark, L. (2019) Education as property, *Virginia Law Review* 105 (2): 397–424.

BBC (2014) Character can and should be taught in schools, says Hunt, *BBC News Online*. Available from: https://www.bbc.co.uk/news/uk-england-london-26140607.

BBC (2019) Felicity Huffman handed prison time over college admissions scandal, *BBC News Online*, 13 September 2019. Available from: https://www.bbc.co.uk/news/world-us-canada-49693193.

Baden-Powell, R. (1922) *Rovering to Success: A Book of Life-Sport for Young Men*. London: Herbert Jenkins.

Baden-Powell, R. (1936) *Adventuring to Manhood*. London: C Arthur Pearson Ltd.

Baden-Powell, R. (1944) [1919] *Aids to Scoutmastership*. World Brotherhood Edition. New York, NY: Boy Scouts of America.

Baden-Powell, R. (2004) [1908] *Scouting for Boys: A Handbook for Instruction in Good Citizenship*. Oxford: Oxford University Press.

Baillie Smith, M. and Laurie, N. (2011) International volunteering and development: Global citizenship and neoliberal professionalisation today, *Transactions of the Institute of British Geographers* 36 (4): 545–559.

Baker, A. R. H. (2017) *Amateur Musical Societies and Sports Clubs in Provincial France, 1848-1914: Harmony and Hostility*. Basingstoke: Palgrave Macmillan.

Barber, M. (1994) *The Making of the 1944 Education Act*. London: Cassell.

Barnett, C. (2012) Geography and ethics, *Progress in Human Geography* 36: 379–388.

Basham, V. M. (2016) Raising an army: The geopolitics of militarizing the lives of working-class boys in an age of austerity, *International Political Sociology* 10 (3): 258–274.

Beaumont, P. (2017) Brexit, retrotopia and the perils of post-colonial delusions, *Global Affairs* 3 (4–5): 379–390.

Bennett, E., Coule, T., Damm, C., Dayson, C., Dean, J. and Macmillan, R. (2019) Civil society strategy: A policy review, *Voluntary Sector Review* 10 (2): 213–223.

Benwell, M. C. and Hopkins, P. E. (eds) (2016) *Children, Young People and Critical Geopolitics*. London: Routledge.

Berkowitz, M. W. (2002) The Science of Character Education, in W. Damon (ed) *Bringing in a New Era of Character Education*. Stanford, CA: Hoover Institution Press: 43–63.

Berkowitz, M. W. and Bier, M. C. (2005) *What Works in Character Education: A Research-Drive Guide for Educators*. Washington, DC: Character Education Partnership.

Bhambra, G. K. (2017) Locating Brexit in the Pragmatics of Race, Citizenship and Empire, in W. Outhwaite (ed) *Brexit: Sociological Responses*. London: Anthem Press: 91–100.

Birdwell, J. R. S., Scott, R. and Reynolds, L. (2015) *Service Nation 2020*. London: Demos. Available from: https://www.demos.co.uk/files/ServiceNation2020.pdf?1436715297.

Blazek, M. (2015) *Rematerialising Children's Agency: Everyday Practices in a Post-Socialist Estate*. Bristol: Policy Press.

Blunkett, D. (2016) The Impact of NCS Will Be Felt in All Parts of Our Society, *Generation Change Blog*. Available from: http://www.generationchange.org.uk/blog-archive6.html.

Body, A., Holman, K. and Hogg, E. (2017) To bridge the gap? Voluntary action in primary schools, *Voluntary Sector Review* 8 (3): 251–271.

Boehmer, E. (2004) Introduction and Notes on Scouting for Boys, in R. Baden-Powell (2004) [1908] *Scouting for Boys*. Oxford: Oxford University Press: xi–xxxix.

Bowlby, S. (2011) Friendship, co-presence and care: Neglected spaces, *Social & Cultural Geography* 12 (6): 605–622.

Breslin, T. (2016) Citizenship, values and the broader social and personal development curriculum, *BERA Research Intelligence* 130 (Summer 2016): 21–22.

Briggs, D. (2012) *The English Riots of 2011: A Summer of Discontent*. Hampshire: Waterside Press.

British Army (2020) *Who Are We: Army Cadet Force*. Available from: https://www.army.mod.uk/who-we-are/the-armys-cadets/.

Brock, C. (2016) *Geography of Education: Scale, Space and Location in the Study of Education*. London: Bloomsbury.

Brock-Utne, B. (2000) *Whose Education For All? The Recolonization of the African Mind*. London; New York, NY: Falmer Press.

Brooks, R., Fuller, A. and Waters, J. (eds) (2012) *Changing Spaces of Education: New Perspectives on the Nature of Learning*. London: Routledge.

Brooks, R. and Waters, J. (2018) *Materialities and Mobilities in Education*. London: Routledge.

Brown, G. (2011) Emotional geographies of young people's aspirations for adult life, *Children's Geographies* 9 (1): 7–22.

Brown, P. (2012) Education, opportunity and the prospects for social mobility, *British Journal of Sociology of Education* 34 (5–6): 678–700.

Bryce, T. and Humes, W. (2003) *Scottish Education: Post-Devolution*. Edinburgh: Edinburgh University Press.

Buckingham, D. (2000) *The Making of Citizens: Young People, News and Politics*. London: Routledge.

Bull, A. (2016) El Sistema as a Bourgeois Social Project: Class, gender, and Victorian values, *Action, Criticism & Theory for Music Education* 15 (1): 120–153.

Bull, A. (2019a) *Submission to UK Government Consultation on Sex and Relationships Education in Schools*, 1 April 2019. Available from: https://annabullresearch.word-press.com/2019/04/01/submission-to-uk-govt-consultation-on-sex-and-relation-ships-education-in-schools/#ConsentHistories.

Bull, A. (2019b) *Class, Control and Classical Music.* Oxford: Oxford University Press.

Bull, A. and Allen, K. (2018) Introduction: Sociological interrogations of the turn to character, *Sociological Research Online* 23 (2): 392–298.

Burman, E. (2018) (Re)sourcing the character and resilience manifesto: Suppressions and slippages of (re)presentation and selective affectivities, *Sociological Research Online* 23 (2): 416–437.

Butler, T. and Hamnett, C. (2010) 'You Take What you are Given': The limits to parental choice in education in East London, *Environment and Planning A* 42: 2431–2450.

BVI News (2019) Gov't to revise Student Code of Conduct policy to 'build character' of youths, BVINews.com, 18 November 2019. Available from: https://bvinews.com/govt-to-revise-student-code-of-conduct-policy-to-build-character-of-youths/.

Cameron, D. (2016) I've found my first job after politics, building the Big Society, *The Telegraph*, 11 October 2016. Available from: https://www.telegraph.co.uk/news/2016/10/11/david-cameron-ive-found-my-first-job-after-politics-building-the/.

Cameron, L. (2006) Science, nature, and hatred: 'Finding Out' at the Malting House Garden School, 1924–29, *Environment and Planning D: Society and Space* 24 (6): 851–872.

Carr, D. (2006) The moral roots of citizenship: Reconciling principle and character in citizenship education, *Journal of Moral Education* 35 (4): 443–456.

Carr, D. (ed) (2016) *Perspectives on Gratitude: An Interdisciplinary Approach.* London: Routledge.

Cartwright, I. (2012) Informal Education in Compulsory Schooling in the UK, in P. Kraftl, J. Horton and F. Tucker (eds) *Critical Geographies of Childhood and Youth: Contemporary Policy and Practice.* Bristol: Policy Press: 151–168.

Castree, N., Kitchen, R. and Rogers, A. (2013) *A Dictionary of Human Geography.* Oxford: Oxford University Press.

Character Lab (2016) *Character Growth Card*, November 2016. Available from: https://www.greatschoolspartnership.org/wp-content/uploads/2016/11/CharacterGrowthCard.pdf.

Chrisafis, A. (2019) Macron's national service sparks criticism from French left. *The Guardian*, 19 June 2019. Available from: https://www.theguardian.com/world/2019/jun/19/rollout-of-compulsory-civic-service-for-young-people-in-france-sparks-criticisms.

Clark, A. (2007) Flying the flag for mainstream Australia. *Griffith Review*, February 2007. Available from: https://www.griffithreview.com/articles/flying-the-flag-for-mainstream-australia/.

Clay, D. and Thomas, A. (2014) *Review of Military Ethos Alternative Provision Projects: Research Report*, December 2014, GOV.UK. Available from: https://www.gov.uk/government/publications/military-ethos-alternative-provision-projects-review Department for Education: London.

Collins, D. and Coleman, T. (2008) Social geographies of education: Looking within, and beyond, school boundaries, *Geography Compass* 2 (1): 281–299.

Conservatives (2007) *It's Time to Inspire Britain's Teenagers: National Citizen Service for the 21st Century: A Six-Week Programme for Every School Leaver.* Available from: http://conservativehome.blogs.com/interviews/files/timetoinspire.pdf.

Cook, L. (1999) The 1944 education act and outdoor education: From policy to practice, *History of Education* 28 (2): 157–172.

Cranston, N., Kimber, M., Mulford, B., Reid, A. and Keating, J. (2010) Politics and school education in Australia: A case of shifting purposes, *Journal of Educational Administration* 48 (2): 182–195.

Cranston, S. (2014) Reflections on doing the expat show: Performing the Global Migration Industry, *Environment and Planning A* 46 (5): 1124–1138.

Cranston, S. (2017) Self-help and the surfacing of identity: Producing the third-culture kid, *Emotion, Space and Society* 24: 27–33.

Cranston, S., Schapendonk, J. and Spaan, E. (2018) New directions in exploring the migration industries: Introduction to special issue, *Journal of Ethnic and Migration Studies* 44 (4): 543–557.

Credé, M. (2018) What shall we do about grit? A critical review of what we know and what we don't know, *Educational Researcher* 47 (9): 606–611.

Cresswell, T. (1996) *In Place/out of Place: Geography, Ideology, and Transgression.* London; Minneapolis: University of Minnesota Press.

Cresswell, T. (2005) Moral Geographies, in D. Atkinson, P. Jackson, D. Sibley and N. Washbourne (eds) *Geography: A Critical Dictionary of Key Concepts.* London; New York, NY: Taurus: 128–134.

Cresswell, T. (2010) New cultural geography – An unfinished project? *Cultural Geographies* 17 (2): 169–174.

Cupers, K. (2008) Governing through nature: Camps and youth movements in inter-war Germany and the United States, *Cultural Geographies* 15 (2): 173–205.

Daddow, O. (2019) GlobalBritain™: The discursive construction of Britain's post-Brexit world role, *Global Affairs* 5 (1): 5–22.

Davies, B. (2017) Youth volunteering: the new panacea? *Youth&Policy Blog.* Available from: https://www.youthandpolicy.org/articles/youth-volunteering-the-new-panacea/.

Davies, B. (2018) *Austerity, Youth Policy and the Deconstruction of the Youth Service in England.* Basingstoke: Palgrave Macmillan.

Davies, R. (2014) After School: The Disruptive Work of Informal Education, in S. Mills and P. Kraftl (eds) (2014) *Informal Education, Childhood and Youth: Geographies, Histories, Practices.* Basingstoke: Palgrave Macmillan: 216–228.

Davies, I., Gorard, S. and McGuinn, N. (2005) Citizenship education and character education: Similarities and contrasts, *British Journal of Educational Studies* 53 (3): 341–358.

Day, C. and Evans, R. (2015) Caring responsibilities, change and transitions in young people's lives in Zambia, *Journal of Comparative Family Studies* 46 (1): 137–152.

DCMS (2016) *Policy Paper: Social Action*, 13 July 2016. Available from: https://www.gov.uk/government/publications/centre-for-social-action/centre-for-social-action.

DCMS (2020) *National Citizen Service 2018 Evaluation*, Kantar and London Economics. Available from: https://www.gov.uk/government/publications/national-citizen-service-evaluation-report-2018.

DCSF (2007) *Social and Emotional Aspects of Learning for Secondary Schools.* Nottingham: Department for Children, Schools and Families Publications.

de Leeuw, S. (2009) 'If anything is to be done with the Indian, we must catch him very young': Colonial constructions of aboriginal children and the geographies of Indian residential schooling in British Columbia, Canada, *Children's Geographies* 7 (2): 123–140.

de Soria, A. B. M., Durána, C. N., Morrása, A. S., Gonzáleza, J. P. D. and Varela, A. G. (2018) Questions and answers regarding character education in Latin American countries (Mexico, Colombia and Argentina): An exploratory Delphi study, *Edetania* 53 (Julio 2018): 23–44. Available from: https://dialnet.unirioja.es/servlet/articulo?codigo=6581949.

de St Croix, T. (2011) Struggles and silences: Policy, youth work and the National Citizen Service. *Youth and Policy* 106: 43–59.

de St Croix, T. (2018) Youth work, performativity and the new youth impact agenda: Getting paid for numbers? *Journal of Education Policy* 33 (3): 414–438.

Demos (2016) Ambitious pilot in deprived schools shows how extra-curricular activities can build children's leadership skills. *Demos*. Available from: https://demosuk.wpengine.com/press-release/character-by-doing-evaluation/.

Demos (2014) *Speech to Character Conference, Tristram Hunt MP*, 8 December 2014, London: Demos. Available from: https://www.demos.co.uk/files/TristramHuntspeech.pdf.

Dewey, J. (1944) *Democracy and Education*. New York, NY: The Free Press.

DfE (2014a) *Measures to Help Schools Instil Character in Pupils Announced*. GOV.UK. Available from: https://www.gov.uk/government/news/measures-to-help-schools-instil-character-in-pupils-announced.

DfE (2014b) *DfE Character Awards Application Window Now Open*. GOV.UK. Available from: https://www.gov.uk/government/news/dfe-character-awards-application-window-now-open.

DfE (2014c) *England to Become a Global Leader of Teaching Character*. GOV.UK. Available from: https://www.gov.uk/government/news/england-to-become-a-global-leader-of-teaching-character.

DfE (2014d) Promoting fundamental British values as part of SMSC in schools: Departmental advice for maintained schools, November 2014. Available from: https://assets.publishing.service.gov.uk/government/uploads/system/uploads/attachment_data/file/380595/SMSC_Guidance_Maintained_Schools.pdf.

DfE (2014e) The National Curriculum in England: Framework Document, December 2014. Available from: https://assets.publishing.service.gov.uk/government/uploads/system/uploads/attachment_data/file/381344/Master_final_national_curriculum_28_Nov.pdf.

DfE (2015a) *DfE Character Awards Application Window Now Open*. GOV.UK. Available from: https://www.gov.uk/government/news/dfe-character-awards-application-window-now-open.

DfE (2015b) *Rugby Coaches to be Drafted in to Help Build Grit in Pupils*. GOV.UK. Available from: https://www.gov.uk/government/news/rugby-coaches-to-be-drafted-in-to-help-build-grit-in-pupils.

DfE (2015c) *The Prevent Duty: Departmental Advice for Schools and Childcare Providers*, June 2015. Available from: https://assets.publishing.service.gov.uk/government/uploads/system/uploads/attachment_data/file/439598/prevent-duty-departmental-advice-v6.pdf.

DfE (2017a) *Developing Character Skills in Schools*, August 2017. Available from: https://assets.publishing.service.gov.uk/government/uploads/system/uploads/attachment_data/file/634710/Developing_Character_skills-synthesis_report.pdf London: NatCen.

DfE (2017b) *National Citizen Service: Guidance for Schools and Colleges*. Available from: https://www.gov.uk/government/publications/national-citizen-service-guidance-for-schools-and-colleges.

DfE (2018) Government invests £5m to increase places for disadvantaged children in youth organisations. GOV.UK. Available from: https://www.gov.uk/government/news/government-invests-5m-to-increase-places-for-disadvantaged-children-in-youth-organisations

DfE (2019a) *Character Education: Framework Guidance*, November 2019. Available from: https://www.gov.uk/government/publications/character-education-framework.

DfE (2019b) Education Secretary sets out five foundations to build character. GOV.UK. Available from: https://www.gov.uk/government/speeches/education-secretary-sets-out-five-foundations-to-build-character.

DfE (2019c) *Personal, Social, Health and Economic (PSHE) Education*. Available from: https://www.gov.uk/government/publications/personal-social-health-and-economic-education-pshe/personal-social-health-and-economic-pshe-education.

DfE (2019d) *Character and Resilience: A Call for Evidence*. GOV.UK. Available from: https://www.gov.uk/government/consultations/character-and-resilience-call-for-evidence.

Dickens, L. and Lonie, D. (2016) Becoming musicians: Situating young people's experiences of musical learning between formal, informal and non-formal spheres, *Cultural Geographies* 23 (1): 87–101.

Divala, J. J. and Enslin, P. (2008) Citizenship Education in Malawi: Prospects for Global Citizenship, in J. Arthur, I. Davies and C. Hahn (eds) *Handbook of Education for Citizenship and Democracy*. London: SAGE: 215–222.

Donnelly, M., Lazetic, P., Sandoval-Hernández, A., Kameshwara, K. K. and Whewall, S. (2019). *An Unequal Playing Field: Extra-Curricular Activities, Soft Skills and Social Mobility*. London: Social Mobility Commission.

Dorling, D. and Tomlinson, S. (2019) *Rule Britannia: Brexit and the End of Empire*. London: Biteback Publishing.

Driver, F. (1988) Moral geographies: Social science and the urban environment in mid-nineteenth century England, *Transactions of the Institute of British Geographers* 13 (3): 275–287.

Duckworth, A. (2016) *GRIT – Why Passion and Resilience Are the Secrets to Success*. London: Vermillion.

Dungey, C. E. and Ansell, N. (2020) 'I go to school to survive': Facing physical, moral and economic uncertainties in rural Lesotho, *Children's Geographies* 18 (6): 614–628.

Dunkley, C. (2009) A therapeutic taskscape: Theorizing place-making, discipline and care at a camp for troubled youth, *Health and Place* 15 (1): 88–96.

Dwyer, C. and Parutis, V. (2012) 'Faith in the system?' State-funded faith schools in England and the contested parameters of community cohesion, *Transactions of the Institute of British Geographers* 38 (2): 267–284.

Edmonson, S., Tatman, R. and Slate, J. R. (2009) Character education: A critical analysis, *International Journal of Educational Leadership Preparation* 4 (4), Oct–Dec 2009. Available from: https://files.eric.ed.gov/fulltext/EJ1068485.pdf.

Edwards, G. (2002) Geography, Culture, Values and Education, in R. Gerber and M. Williams (eds) *Geography, Culture and Education*. Dordrecht: Kluwer: 31–40.

Edwards, S. (2018) *Youth Movements, Citizenship and the English Countryside: Creating Good Citizens 1930-1960*. London: Palgrave Macmillan.

Enslin, P. and Horsthemke, K. (2004) Can *Ubuntu* provide a model for citizenship education in African democracies? *Comparative Education* 40 (4): 545–558.

Etieyibo, E. (2017) Moral education, Ubuntu and Ubuntu-inspired communities, *South African Journal of Philosophy* 36 (3): 311–325.

Facer, K. (2011) *Learning Futures: Education, Technology and Social Change*. London; New York, NY: Routledge.

Fairless Nicholson, J. (2020, online early view). From London to Grenada and Back Again: Youth Exchange Geographies and the Grenadian Revolution, 1979–1983, *Antipode*. DOI: https://doi.org/10.1111/anti.12598.

Feinberg, W. (2006) *For Goodness Sake: Religious Schools and Education for Democratic Citizenry. London*; New York, NY: Routledge.

Fesmire, S. (2003) *John Dewey and Moral Imagination: Pragmatism in Ethics.* Bloomington, IN: Indiana University Press.

Finn, M. (2016) Atmospheres of progress in a data-based school, *Cultural Geographies* 23 (1): 29–49.

Freeman, M. (2011) From 'character-training' to 'personal growth': The early history of the outward bound 1941-1965, *History of Education* 40 (1): 21–43.

Freeden, M. (2003) Civil Society and the Good Citizen: Competing Conceptions of Citizenship in Twentieth-century Britain, in J. Harris (ed) *Civil Society in British History: Ideas, Identities, Institutions.* Oxford: Oxford University Press, 275–291.

Friedman, S. and Laurison, D. (2019) *The Class Ceiling: Why It Pays to Be Privileged.* Bristol: Policy Press.

Gagen, E. A. (2000) An example to us all: Child development and identity construction in early 20th-century playgrounds, *Environment and Planning A* 32 (4): 599–616.

Gagen, E. A. (2004a) Making America flesh: Physicality and nationhood in early twentieth century physical education reform, *Cultural Geographies* 11 (4): 417–442.

Gagen, E. A. (2004b) Landscapes of Childhood and Youth, in J. S. Duncan, N. Johnson and R. H. Schein (eds) *A Companion to Cultural Geography*. Oxford: Blackwell: 404–419.

Gagen, E. A. (2015) Governing emotions: Citizenship, neuroscience and the education of youth, *Transactions of the Institute of British Geographers* 40 (1): 140–152.

Gamsu, S. (2018) The 'other' London effect: The diversification of London's suburban grammar schools and The rise of hyper-selective elite state schools, *The British Journal of Sociology* 69 (4): 1155–1174.

Gawlik, M. A. (2016) The US Charter School Landscape: Extant literature, gaps in research, and implications for the US educational system, *Global Education Review* 3 (2): 50–83.

George, M. (2017) DfE scraps Nicky Morgan's 'landmark' character education scheme. *TES*. Available from: https://www.tes.com/news/dfe-scraps-nicky-morgans-landmark-character-education-scheme.

Gill, R. and Orgad, S. (2018) The amazing bounce-backable woman: Resilience and the psychological turn in neoliberalism, *Sociological Research Online* 23 (2): 477–495.

Goldstein, D. (2015) Inexcusable Absences, 6 March 2015. Available from: https://newrepublic.com/article/121186/truancy-laws-unfairly-attack-poor-children-and-parents.

Goodwin, M., Jones, M. and Jones, R. (2005) Devolution, constitutional change and economic development: Explaining and understanding the new institutional geographies of the British state, *Regional Studies* 39 (4): 421–436.

Gorard, S. and Siddiqui, N. (2018) Grammar schools in England: A new analysis of social segregation and academic outcomes, *British Journal of Sociology of Education* 39 (7): 909–924.

Griffin, C. (2001) Imagining new narratives of youth: Youth research, the 'New Europe' and global youth culture, *Childhood* 8 (2): 147–166.

Gruffudd, P. (1996) The countryside as educator: Schools, rurality and citizenship in inter-war Wales, *Journal of Historical Geography* 22 (4): 412–423.

Gulf News (2019) Schools to roll out standardised tests in Moral Education, *Gulf News*, 13 April 2019. Available from: https://gulfnews.com/uae/education/schools-to-roll-out-standardised-tests-in-moral-education-1.63310655.

Gutman, M. and de Coninck-Smith, N. (2008) *Designing Modern Childhoods: History, Space, and the Material Culture of Children*. New Brunswick, NJ: Rutgers University Press.

Hall, S. M. (2019) Everyday austerity: Towards relational geographies of family, friendship and intimacy, *Progress in Human Geography* 43 (5): 769–789.

Hanson Thiem, C. (2009) Thinking through education: The geographies of contemporary educational restructuring, *Progress in Human Geography* 33 (2): 154–173.

Haque, Z. (2019) The Home Office's harsh 'good character' tests are ruining children's lives. *The Guardian*, 10 July 2019. Available from: https://www.theguardian.com/commentisfree/2019/jul/10/home-office-good-character-test-children-british-citizenship.

Harrison, T. and Bawden, M. (2016) Teaching character through subjects, *BERA Research Intelligence* 130 (Summer 2016): 5–16.

Hatcher, R. (2011) The Conservative-Liberal Democrat Coalition government's "free schools" in England, *Educational Review* 63 (4): 485–503.

Hazlewood, R. and Thurman, J. (1950) *The Camp Fire Leader's Book*. London: Boy Scout Association.

Heater, D. (2001) The history of citizenship education in England, *The Curriculum Journal* 12 (1): 103–123.

Hemming, P. J. (2015) *Religion in the Primary School: Ethos, Diversity, Citizenship*. London: Routledge.

Herod, A. (2009) *Scale*. London: Routledge.

Hickman Dunne, J. (2019) Experiencing the outdoors: Embodied encounters in the Outward Bound Trust, *The Geographical Journal* 185 (3): 279–291.

Hickman Dunne, J. and Mills, S. (2019) Educational landscapes: Nature, place and moral geographies, *The Geographical Journal* 185 (3): 254–257.

Hinds, D. (2019) I want every child to have the perseverance, grit and determination of the Lionesses, *The Telegraph*. https://www.telegraph.co.uk/world-cup/2019/06/21/want-every-child-have-perseverance-grit-determination-lionesses/.

HM Government (2017) *The Shared Society: Prime Minister's Speech at the Charity Commission Annual Meeting*, 9 January 2017. Available from: https://www.gov.uk/government/speeches/the-shared-society-prime-ministers-speech-at-the-charity-commission-annual-meeting.

Holdsworth, C. (2017) The cult of experience: Standing out from the crowd in an era of austerity, *Area* 49 (3): 296–302.

Holloway, S. L. (2014) Changing children's geographies, *Children's Geographies* 12 (4): 377–392.

Holloway, S. L., Hubbard, P., Jöns, H. and Pimlott-Wilson, H. (2010) Geographies of education and the significance of children, youth and families, *Progress in Human Geography* 34 (5): 583–600.

Holloway, S. L. and Jöns, H. (2012) Geographies of education and learning, *Transactions of the Institute of British Geographers* 37 (4): 482–488.

Holloway, S. L. and Kirby, P. (2020) Neoliberalising education: New geographies of private tuition, class privilege and minority ethnic advancement, *Antipode* 52 (1): 164–184.

Holloway, S. L. and Pimlott-Wilson, H. (2011) The politics of aspiration: Neo-liberal education policy, 'low' parental aspirations, and primary school extended services in disadvantaged communities. *Children's Geographies* 9 (1): 79–94.

Holloway, S. L. and Pimlott-Wilson, H. (2014a) Enriching children, institutionalizing childhood? Geographies of play, extracurricular activities, and parenting in England, *Annals of the Association of American Geographers* 104 (3): 613–627.

Holloway, S. L. and Pimlott-Wilson, H. (2014b) "Any Advice is Welcome Isn't it?": Neoliberal parenting education, local mothering cultures, and social class, *Environment and Planning A* 46 (1): 94–111.

Holloway, S. L. and Valentine, G. (eds) (2000) *Children's Geographies: Playing, Living, Learning*. London: Routledge.

Holt, L. (2011) *Geographies of Children, Youth and Families*. London: Routledge.

Holt, L., Bowlby, S. and Lea, J. (2017) "Everyone knows me…. I sort of like move about": The friendships and encounters of young people with special educational needs in different school settings, *Environment and Planning A: Economy and Space* 49 (6): 1361–1378.

Home Office (2019) *Nationality: Good Character Requirement*, Version 1. Published for Home Office staff on 14 January 2019. Available from: https://assets.publishing.service.gov.uk/government/uploads/system/uploads/attachment_data/file/770960/good-character-guidance.pdf.

Hopkins, P. E. (2007) *Young People, Place and Identity*. London: Routledge.

Horton, J. and Kraftl, P. (2013) *Cultural Geographies: An Introduction*. London: Routledge.

House of Lords (2018) *The Ties That Bind: Citizenship and Civic Engagement in the 21st Century*. HL 2017-2019 [118], London: House of Lords.

Howard, R. W., Berkowitz, M. W. and Schaeffer, E. F. (2004) Politics of character education, *Educational Policy* 18 (1): 188–215.

Hubery, D. S. (1963) *The Emancipation of Youth*. London: The Epworth Press.

Inkeles, A. (1996) *National Character: A Psycho-Social Perspective*. London: Routledge.

Inspira (2018) *Volunteering Hours Continue to Rise with NCS*. Available from: https://www.inspira.org.uk/news/volunteering-hours-continue-to-rise-with-ncs.

IPEN (2020) International Positive Education Network Website. Available from: http://www.ipen-network.com/.

Isin, E. F. and Wood, P. K. (1999) *Citizenship and Identity*. London: SAGE.

Ismay, J. (2019) Pete Buttigieg Proposes National Service Programs for Climate Change and Mental Health, *New York Times*. Available from: https://www.nytimes.com/2019/07/03/us/politics/buttigieg-national-service.html.

Jackson, P. (1989) *Maps of Meaning: An Introduction to Cultural Geography*. London; New York, NY: Routledge.

Jarvis, D. (2016) Thousands of Young People Doing Their Bit for Society? Let's Be More Ambitious, *The Guardian*, 13 April 2016. Available from: https://www.theguardian.com/commentisfree/2016/apr/13/young-people-national-citizen-service-sense-of-purpose-community.

Jeal, T. (1989) *Baden-Powell: Founder of the Boy Scouts*. New Haven, CT and London: Yale University Press.

Jeffrey, C. (2010) Geographies of children and youth I: Eroding maps of life, *Progress in Human Geography* 34 (4): 496–505.

Jeffs, T. and Smith, M. (2005) *Informal Education: Conversation, Democracy and Learning*. London: Educational Heretics Press.

Jenks, C. (1996) *Childhood*. London: Routledge.

Jerome, L. and Kisby, B. (2019) *The Rise of Character Education in Britain: Heroes, Dragons and the Myths of Character*. Cham, Switzerland: Palgrave Macmillan.

Johnson-Hanks, J. (2002) On the limits of life stages in ethnography: Toward a theory of vital conjunctures, *American Anthropologist* 104 (3): 865–880.

Jones, T. M. (2009) Framing the framework: Discourses in Australia's national values education policy, *Educational Research for Policy and Practice* 8 (1): 35–57.

Jöns, H., Meusburger, P. and Heffernan, M. (2017) *Mobilities of Knowledge*. Berlin: Springer.

Kadish, S. (1995) *'A Good Jew and a Good Englishman': The Jewish Lads' & Girls' Brigade, 1895-1995*. London: Valentine Mitchell.

Kallio, K. P. and Häkli, J. (2011) Are there politics in childhood? *Space and Polity* 15 (1): 21–34.

Kallio, K. P. and Häkli, J. (2015) Children's Political Geographies, in J. Agnew, V. Mamadouh, A. Secor and J. Sharp (eds) *The Wiley-Blackwell Companion to Political Geography*. Hoboken, NJ: Wiley-Blackwell: 265–278.

Katz, C. (2008) Childhood as spectacle: Relays of anxiety and the reconfiguration of the child, *Cultural Geographies* 15 (1): 5–17.

Kearns, A. (1992) Active citizenship and urban governance, *Transactions of the Institute of British Geographers* 17 (1): 20–34.

Kearns, A. (1995) Active citizenship and local governance: Political and geographical dimensions, *Political Geography* 14 (2): 155–75.

Keating, A., Kerr, D., Benton, T., Mundy, E. and Lopes, J. (2010) *Citizenship Education in England 2011-2010: Young People's Practices and Prospects for the Future: The Eighth and Final Report from the Citizenship Education Longitudinal Study (CELS)*. Department for Education, Research Report DFE-RR059, ISBN: 978-1-84775-822-4.

Kennelly, J. and Llewellyn, K. R. (2011) Educating for active compliance: Discursive constructions in citizenship education, *Citizenship Studies* 15 (6–7): 897–914.

Kerr, D. (1999) Citizenship education in the curriculum: An international review, *The School Field X* (3/4): 5–32.

Klaiber, J. (2009) The Catholic Church, moral education and citizenship in Latin America, *Journal of Moral Education* 38 (4): 407–420.

Knapton, S. and Hope, C. (2019) Every child in Britain must spend a night under the stars as too many have lost touch with nature, government warned, *The Telegraph*. Available from: https://www.telegraph.co.uk/science/2019/09/20/every-child-britain-must-spend-night-stars-many-have-lost-touch/.

Kohn, A. (1997) How not to teach values: A critical look at character education, *Phi Beta Kappa* 78 (6): 429–439.

Kraftl, P. (2006) Building an idea: The material construction of an ideal childhood, *Transactions of the Institute of British Geographers* 31 (4): 488–504.

Kraftl, P. (2010) Events of Hope and Events of Crisis: Young People and Hope in the UK, in J. Leaman and M. Woersching (eds) *Youth in Contemporary Europe*. London: Routledge: 215–231.

Kraftl, P. (2013) *Geographies of Alternative Education: Diverse Learning Spaces for Children and Young People*. Bristol: Policy Press.

Kraftl, P., Horton, J. and Tucker, F. (eds) (2012) *Critical Geographies of Childhood and Youth: Contemporary Policy and Practice*. Bristol: Policy Press.

Kristjánsson, K. (2013) Ten myths about character, virtue and virtue education – Plus three well-founded misgivings, *British Journal of Educational Studies* 61 (3): 269–287.

Kristjánsson, K. (2015) *Aristotelian Character Education*. London: Routledge.

Kuo, L. (2019) 'Defend China's honour': Beijing releases new morality guidelines for citizens, *The Guardian*, 29 October 2019. Available from: https://www.theguardian.com/world/2019/oct/29/defend-chinas-honour-beijing-releases-new-morality-guidelines-for-citizens.

Kurohi, R. (2019) MOE to review character and citizenship education syllabus to focus on moral education for younger pupils, *Straits Times*, 8 November 2019. Available from: https://www.straitstimes.com/singapore/moe-to-review-character-and-citizenship-education-syllabus-to-focus-on-moral-education-for.

Kyle, R. G. (2007) *Moral Geographies of the Boys' Brigade in Scotland*, Unpublished PhD Thesis, University of Glasgow.

Lamb, M., Taylor-Collins, E. and Silvergate, C. (2019) Character education for social action: A conceptual analysis of the #iwill campaign, *Journal of Social Science Education* 18(1): https://doi.org/10.4119/jsse-918.

Lander, V. (2016) Introduction to fundamental British values, *Journal of Education for Teaching* 42 (3): 274–279.

Langevang, T. (2007) Movements in time and space: Using multiple methods in research with young people in Accra, Ghana, *Children's Geographies* 5 (3): 267–282.

Lee, Chi-Ming, (2007) Changes and challenges for moral education in Taiwan, *Journal of Moral Education* 33 (4): 575–595.

Legg, S. and Brown, M. (2013) Moral regulation: Historical geography and scale, *Journal of Historical Geography* 42: 134–139.

Letseka, M. (2012) In defence of *Ubuntu*, *Studies in Philosophy and Education* 31: 47–60.

Lickona, T. (1992) *Educating for Character: How Our Schools Can Teach Respect and Responsibility*. New York, NY: Bantam.

Lister, R., Smith, N., Middleton, S. and Cox, L. (2003) Young people talk about citizenship: Empirical perspectives on theoretical and political debates, *Citizenship Studies* 7 (2): 235–253.

Livingstone, S. and Blum-Ross, A. (2020) *Parenting for a Digital Future: How Hopes and Fears About Technology Shape Children's Lives*. New York, NY: Oxford University Press.

Local Government Association (2018) *LGA – National Citizen Service funding should be devolved to local youth services*. Available from: https://www.local.gov.uk/about/news/lga-national-citizen-service-funding-should-be-devolved-local-youth-services.

Louv, R. (2005) *Last Child in the Woods: Saving Our Children from Nature-Deficit Disorder*. New York, NY: Algonquin Books.

Love, B. L. (2019a) *We Want to Do More Than Survive: Abolitionist Teaching and the Pursuit of Educational Freedom*. Boston, MA: Beacon Press.

Love, B. L. (2019b) 'Grit is in our DNA': Why Teaching Grit is Inherently Anti-Black, *Education Week*, 12 February 2019. Available from: https://www.edweek.org/leadership/opinion-grit-is-in-our-dna-why-teaching-grit-is-inherently-anti-black/2019/02.

MacCunn, J. (1931) [1900] *The Making of Character: Some Educational Aspects of Ethics*. Cambridge: Cambridge University Press.

Malik, K. (2020) Shamima Begum is British and her citizenship should not be revoked. *The Guardian*, 9 February 2020. Available from: https://www.theguardian.com/commentisfree/2020/feb/09/britain-has-moral-duty-to-bring-back-shamina-begum.

Manchester, H. and Bragg, S. (2013) School ethos and the spatial turn: "Capacious" approaches to research and practice, *Qualitative Inquiry* 19 (10): 818–827.

Marshall, T. H. (1950) *Citizenship and Social Class and Other Essays*. Cambridge: Cambridge University Press.

Masudi, F. and Zaman, S. (2019) Schools must teach social, emotional intelligence to children, *Gulf News*. Available from: https://gulfnews.com/uae/education/schools-must-teach-social-emotional-intelligence-to-children-1.66309340.

Mandler, P. (2007) *The English National Character: The History of an Idea from Edmund Burke to Tony Blair*. New Haven, CT: Yale University Press.

Mathews, J. (2020) Why nation's biggest charter network dumped its slogan, 'Work hard. Be nice.' *Washington Post*, 7 July 2020. Available from: https://www.washingtonpost.com/local/education/why-nations-biggest-charter-network-dumped-its-slogan-work-hard-be-nice/2020/07/07/e7896c0a-bf9e-11ea-b4f6-cb39cd8940fb_story.html.

Matless, D. (1994) Moral geography in broadland, *Ecumene* 1 (2): 127–155.

Matless, D. (1995) The Art of Right Living: Landscape and Citizenship, 1918–39, in N. Thrift and S. Pile (eds) *Mapping the Subject: Geographies of Cultural Transformation*. London: Routledge: 93–122.

Matless, D. (1997) Moral geographies of English landscape, *Landscape Research* 22 (2): 141–155.

Matless, D. (1998) *Landscape and Englishness*. London: Reaktion.

Matthews, H. (2001) Citizenship, youth councils and young people's participation, *Journal of Youth Studies* 4 (3): 299–318.

McClellan, B. E. (1999) *Moral Education in America: Schools and the Shaping of Character from Colonial Times to the Present*. New York, NY: Teachers College Press.

McCulloch, G. (2011) *Educational Reconstruction: The 1944 Education Act and the Twenty-First Century*. London: Routledge.

McInerney, L. (2019) A child in a bedsit has more 'character' than your braying public schoolboy, Mr Hinds. *The Guardian*, 19 February. Available from: https://www.theguardian.com/education/2019/feb/19/child-bedsit-more-character-than-braying-public-schoolboy-mr-hinds-education-secretary.

McKendrick, J., Kraftl, P., Mills, S., Gregorius, S. and Sykes, G. (2015) Complex geographies of play provision dis/investment across the UK, *International Journal of Play* 4 (3): 228–235.

Meng-chuan, S. and Hsiao, S. (2019) Character education important for children. *Taipei Times*, 29 October 2019. Available from: http://www.taipeitimes.com/News/taiwan/archives/2019/10/29/2003724846.

Merrifield, R. (2018) Headteacher backs plans to encourage pupils to climb trees, *Worcester News*, 27 November 2018. Available from: https://www.worcesternews.co.uk/news/17256585.headteacher-backs-plans-to-encourage-pupils-to-climb-trees/.

Merriman, P. (2005) 'Respect the life of the countryside': The county code, government and the conduct of visitors to the countryside in post-war England and Wales, *Transactions of the Institute of British Geographers* 30 (3): 336–350.

Metcalfe, J. and Moulin-Stozek, D. (2020) Religious education teachers' perspectives on character education, *British Journal of Religious Education*. DOI: 10.1080/01416200.2020.1713049.

Millar, F. (2015) A nod towards 'character education' is welcome – just don't start measuring it. *The Guardian*, 10 March. Available from: https://www.theguardian.com/education/2015/mar/10/character-education-schools.

Mills, S. (2011a) Scouting for girls? Gender and the scout movement in Britain, *Gender, Place and Culture* 18 (4): 537–556.

Mills, S. (2011b) Be prepared: Communism and the politics of scouting in 1950s Britain, *Contemporary British History* 25 (3): 429–450.

Mills, S. (2012) Duty to God/Dharma/Allah/Waheguru: Diverse youthful religiosities and the politics and performance of informal worship, *Social & Cultural Geography* 13 (5): 481–499.

Mills, S. (2013) 'An instruction in good citizenship': Scouting and the historical geographies of citizenship education, *Transactions of the Institute of British Geographers* 38 (1): 120–134.

Mills, S. (2014) Youth on streets and bob-a-job week: Urban geographies of masculinity, risk and constructions of home in post-war Britain, *Environment and Planning A* 46 (1): 112–128.

Mills, S. (2015) Geographies of youth work, volunteering and employment: The Jewish Lads' Brigade & Club in post-war Manchester, *Transactions of the Institute of British Geographers* 40 (4): 523–535.

Mills, S. (2016) Geographies of education, volunteering and the lifecourse: The woodcraft folk in Britain (1925–1975), *Cultural Geographies* 23 (1): 103–119.

Mills, S. and Kraftl, P. (2014) *Informal Education, Childhood and Youth: Geographies, Histories, Practices*. Basingstoke: Palgrave Macmillan.

Mills, S. and Kraftl, P. (2016) Cultural geographies of education, *Cultural Geographies* 23 (1): 19–27.

Mills, S. and Waite, C. (2017) Brands of youth citizenship and the politics of scale: National Citizen Service in the United Kingdom, *Political Geography* 56 (1): 66–76.

Mills, S. and Waite, C. (2018) From Big Society to shared society? Geographies of social cohesion and encounter in the UK's National Citizen Service, *Geografiska Annaler: Series B Human Geography* 100 (2): 131–148.

Mitchell, K. (2018) *Making Workers: Radical Geographies of Education*. London: Pluto Press.

MOE (2020) *Primary School Education: Preparing Your Child for Tomorrow. Singapore*: Ministry of Education. Available from: https://www.moe.gov.sg/docs/default-source/document/education/primary/files/primary-school-education-booklet.

Mohan, J. (1994) What can you do for your country? Arguments for and against Clinton's National Service legislation, *Policy & Politics* 22 (4): 257–266.

Mohan, J. (2012) Geographical foundations of the Big Society, *Environment and Planning A* 44 (5): 1121–1129.

Morgan, N. (2017) *Taught Not Caught: Educating for 21st Century Character. Woodbridge:* John Catt Educational Ltd.

Morrin, K. (2018) (Re)sourcing the character and resilience manifesto: Suppressions and slippages of (re)presentation and selective affectivities, *Sociological Research Online* 23 (2): 459–476.

Moss, S. M. (2012) *Natural Childhood*. London: The National Trust.

Murray, J. (2010) Students hit by scrapping of education maintenance allowance, *The Guardian*, 25 October 2010. Available from: https://www.theguardian.com/education/2010/oct/25/education-maintenance-allowance.

Mycock, A. and Tonge, J. (2011a) The party politics of youth citizenship and democratic engagement, *Parliamentary Affairs* 65 (1): 138–161.

Mycock, A. and Tonge, J. (2011b) A Big Idea for the Big Society? The advent of National Citizen Service, *The Political Quarterly* 82 (1): 56–66.

Myong, E. (2019) Democratic candidate Pete Buttigieg calls for national service plan that offers student debt relief. *CNBC Online*. 3 July 2019. Available from: https://www.cnbc.com/2019/07/03/pete-buttigieg-calls-for-national-service-plan-with-student-debt-relief.html.

National Audit Office (2017) *Report by the Comptroller and Auditor General: Cabinet Office and Department of Culture, Media & Sport – National Citizen Service* HC916, Session 2016–17, 13 January 2017. Available from: https://www.nao.org.uk/wp-content/uploads/2017/02/National-Citizen-Service.pdf.

National Youth Agency (2019) *Youth Work Inquiry: Final Report Including Recommendations and Summary,* April 2019, APPG on Youth Affairs. Available from: https://nya.org.uk/appg-inquiry-final-report/.

National Youth Agency (2020) *Young People and Social Action*. Available from: https://nya.org.uk/work-with-us/young-people-social-action/.

NCS (2012) National Citizen Service Leaflet. Available from: https://assets.publishing.service.gov.uk/government/uploads/system/uploads/attachment_data/file/88205/NCS-yes_leaflet_no-folder.pdf.

NCS Trust (2019) NCS Trust 2.0 Network Update. Available from: https://www.ncsyes.co.uk/NCS-2.0-network-update.

NCS Trust (2020) One Million Hours of Doing Good. Available from: https://wearencs.com/blog/one-million-hours-doing-good.

Nieto, D. (2018) Citizenship education discourses in Latin America: Multilateral institutions and the decolonial challenge, *Compare: A Journal of Comparative and International Education* 48 (3): 432–450.

Novotný, P., Zimová, E., Mazouchová, A. and Šorgo, A. (2020) Are children actually losing contact with nature, or is it that their experiences differ from those of 120 years ago? *Environment and Behavior*. https://doi.org/10.1177/0013916520937457.

Nucci, L. P. and Narvaez, D. (eds) (2008) *Handbook of Moral and Character Education*. New York, NY: Routledge.

Nugent, R. (2006) Civic, social and political education: Active learning, participation and engagement? *Irish Educational Studies* 25 (2): 207–229.

O'Brien, J. (2020) Are your school values more than just slogans? *TES*, 22 July 2020. Available from: https://www.tes.com/news/are-your-school-values-more-just-slogans.

O'Hara, M. (2014) *Austerity Bites: A Journey into the Sharp End of Cuts in the UK*. Bristol: Policy Press.

Ofsted (2019) *School Inspection Handbook*. Available from: https://assets.publishing.service.gov.uk/government/uploads/system/uploads/attachment_data/file/843108/School_inspection_handbook_-_section_5.pdf.

Osgerby, B. (1998) *Youth in Britain since 1945*. Oxford: Blackwell.

Osler, A. (2016) 'Citizenship education, social justice and Brexit', *BERA Research Intelligence* 130(Summer 2016): 12–14.

Osler, A. and Starkey, H. (2005) *Changing Citizenship: Democracy and Inclusion in Education*. Maidenhead: Open University Press.

Painter, J. (2012) The politics of the neighbour, *Environment and Planning D: Society and Space* 30 (3): 515–533.

Parker, G. (2006) The country code and the ordering of countryside citizenship, *Journal of Rural Studies* 22 (1): 1–16.

Pascal, C. and Bertram, T. (2020) *The Scouts Early Years Programme Evaluation: Final Report*. Birmingham: Centre for Research in Early Childhood.

Paterson, C., Tyler, C. and Lexmond, J. (2014) *Character and Resilience Manifesto.* The All Party Parliamentary Group on Social Mobility with CentreForum and Character Counts. Available from: http://www.educationengland.org.uk/documents/pdfs/2014-appg-social-mobility.pdf.

Percy-Smith, B. (2010) Councils, consultations and community: Rethinking the spaces for children and young people's participation, *Children's Geographies* 8 (2): 107–122.

Peterson, A., Lexmond, J., Hallgarten, J. and Kerr, D. (2014) *Schools with Soul: A New Approach to Spiritual, Moral, Social and Cultural Education (SMSC): Executive Summary* London: RSA. Available from: file:///C:/Users/gysm4/Downloads/Schools-with-soul-a-new-approach-to-spiritual-moral-social-and-cultural-education-smsc-executive-summary.pdf.

Phillips, R. (1997) *Mapping Men and Empire: A Geography of Adventure.* London: Routledge.

Phillips, R. (2003) Education policy, comprehensive schooling and devolution in the disUnited Kingdom: An historical 'home international' analysis, *Journal of Education Policy* 18 (1): 1–17.

Philo, C. and Parr, H. (2000) Institutional geographies: Introductory remarks, *Geoforum* 31 (4): 513–521.

Philo, C. and Smith, F. M. (2003) Guest editorial: Political geographies of children and young people, *Space and Polity* 7 (2): 99–115.

Pietig, J. (2006) John Dewey and character education, *Journal of Moral Education* 6 (3): 170–180.

Pimlott-Wilson, H. (2017) Individualising the future: The emotional geographies of neoliberal governance in young people's aspirations, *Area* 49 (3): 288–295.

Pimlott-Wilson, H. and Coates, J. (2019) Rethinking learning?: Challenging and accommodating neoliberal educational agenda in the integration of forest school into mainstream educational settings, *The Geographical Journal* 185 (3): 268–278.

Ploszajska, T. (1994) Moral landscapes and manipulated spaces: Gender, class and space in Victorian reformatory schools, *Journal of Historical Geography* 20 (4): 413–429.

Ploszajska, T. (1996) Constructing the subject: Geographical models in English schools, 1870–1944, *Journal of Historical Geography* 22 (4): 388–398.

Ploszajska, T. (1998) Down to earth? Geography fieldwork in English schools, 1870–1944, *Environment and Planning D: Society and Space* 16 (6): 757–774.

Ploszajska, T. (2000) Historiographies of geography and empire, in B. Graham and C. Nash (eds) *Modern Historical Geographies.* London: Prentice Hall: 121–145.

Pratt, M. L. (1992) *Imperial Eyes: Travel Writing and Transculturation.* London; New York, NY: Routledge.

Preston, R. (2018) Scouts announces pilot programmes for younger children, 14 November 2018. Available from: https://www.civilsociety.co.uk/news/scouts-to-trial-early-years-services-for-first-time.html.

Prior, R. (2019) Presidential candidates want to expand national service. Here's what that means. *CNN Politics*, 3 July 2019. Available from: https://edition.cnn.com/2019/07/03/politics/national-service-explainer-trnd/index.html.

Proctor, T. M. (2002) *On My Honour: Guides and Scouts in Interwar Britain.* Philadelphia, PA: American Philosophical Society.

Prynn, D. (1983) The Woodcraft Folk and the Labour Movement 1925–70, *Journal of Contemporary History* 18 (1): 79–95.

PSHE Association (2020) Why PSHE Matters. Available from: https://www.pshe-association.org.uk/what-we-do/why-pshe-matters.

Purcell, K. (2011) Discourses of aspiration, opportunity and attainment: Promoting and contesting the academy schools programme, *Children's Geographies* 9 (1): 49–61.

Putnam, R. D. (1995) Bowling alone: America's declining social capital, *Journal of Democracy* 6 (1): 65–78.

Pykett, J. (2009) Making citizens in the classroom. An urban geography of citizenship education? *Urban Studies* 46 (4): 803–823.

Pykett, J. (2011) *Governing Through Pedagogy: Re-Educating Citizens*. London: Routledge.

Pykett, J. (2015) *Brain Culture: Shaping Policy through Neuroscience*. Bristol: Policy Press.

Pykett, J. and Enright, B. (2016) Geographies of brain culture: Optimism and optimisation in workplace training programmes, *Cultural Geographies* 23 (1): 51–68.

Pykett, J., Saward, M. and Schaefer, A. (2010) Framing the good citizen, *The British Journal of Politics and International Relations* 12 (4): 523–538.

QCA (1998) *Education for Citizenship and the Teaching of Democracy in Schools: Final Report of the Advisory Group on Citizenship*. 22 September 1998. London: Qualifications and Curriculum Authority.

Reay, D. (2017) *Miseducation: Inequality, Education and the Working Class*. Bristol: Policy Press.

Rickett, O. (2020) The children who can't afford to be British, *Open Democracy*, 16 January 2020. Available from: https://www.opendemocracy.net/en/openjustice/unlawful-state/the-children-who-cant-afford-to-be-british/.

Ridge, T. (2013) 'We are all in this Together'? The hidden costs of poverty, recession and austerity policies on Britain's poorest children, *Children & Society* 27: 406–417.

Riots Communities and Victims Panel (2012) *After the riots: the final report of the Riots Communities and Victims Panel*, 1 March 2012. Available from: https://www.bl.uk/collection-items/after-the-riots-the-final-report-of-the-riots-communities-and-victims-panel.

Roach, P. (2016) Teaching character: The contribution of teachers to character education, *BERA Research Intelligence* 130 (Summer 2016): 23–24.

Roche, J., Tucker, S., Thomson, R. and Flynn, R. (eds) (2004) *Youth in Society: Contemporary Theory, Policy and Practice London*. Thousand Oaks, CA: SAGE.

Rodriguez, B. (2019). Buttigieg to roll out national service policy for volunteerism. It includes a 'Climate Corps', *USA Today*, 11 December 2019. Available from: https://eu.usatoday.com/story/news/politics/2019/07/03/pete-buttigieg-national-service-policy-proposal-volunteer-climate-corps/1637867001/.

Roesgaard, M. H. (2017) *Moral Education in Japan: Values in a Global Context*. London: Routledge.

Round Square. (2020) *Being Round Square – How*. Available from: https://www.roundsquare.org/being-round-square/how/.

RSA (2019) What role can schools play in developing young people's social agency? *RSA Academies Website*. https://www.rsaacademies.org.uk/what-role-can-schools-play-in-developing-young-peoples-social-agency/.

Russell, W. and Stenning, A. (2020) Beyond active travel: Children, play and community on streets during and after the coronavirus lockdown, *Cities & Health*. https://doi.org/10.1080/23748834.2020.1795386.

Saltman, K. J. (2014) The austerity school: Grit, character and the privatization of public education, *Symploke* 22 (1–2): 41–57.

Sawer, P. (2020) Failings of founder are a lesson, says Chief Scout Bear Grylls, *The Telegraph*, 14 June 2020. Available from: https://www.telegraph.co.uk/news/2020/06/14/failings-founder-lesson-says-chief-scoutbear-grylls/.

Sayer, A. (2020) Critiquing – and rescuing – 'Character', *Sociology* 54 (3): 460–481.

Scott, J. (ed) (2004) *A Dictionary of Sociology*, 4th Edition. Oxford: Oxford University Press.

Scott, R., Reynolds, L. and Cadywould, C. (2016) *Character by Doing: Evaluation. Giving Schools and Non-Formal Learning Providers the Confidence to Work in Partnership*. London: Demos.

Scottish Government (2019) *What Is Curriculum for Excellence?* Available from: https://education.gov.scot/education-scotland/scottish-education-system/policy-for-scottish-education/policy-drivers/cfe-building-from-the-statement-appendix-incl-btc1-5/what-is-curriculum-for-excellence.

Semela, T., Bohl, T. and Kleinknecht, M. (2013) Civic education in Ethiopian schools: Adopted paradigms, instructional technology, and democratic citizenship in a multicultural context, *International Journal of Educational Development* 33: 156–164.

Sigauke, A. T. (2012) Young people, citizenship and citizenship education in Zimbabwe, *International Journal of Educational Development* 32: 214–223.

Singh, A. (2020) Hereditary Tory Peer Suggests 'Grit and Perseverance' More Important Than A-Levels, 14 August 2020. Available from: https://www.huffingtonpost.co.uk/entry/a-levels-lord-bethell-tory_uk_5f367d4dc5b65bbd8c8be966.

Skelton, T. (2010) Taking young people as political actors seriously: Opening the borders of political geography, *Area* 42 (2): 145–151.

Skelton, T. (2013) Children, young people, politics and space: A decade of youthful political geography scholarship 2003–2013, *Space and Polity* 17 (1): 123–136.

Skelton, T. and Valentine, G. (eds) (1998) *Cool Places: Geographies of Youth Cultures*. London: Routledge.

Sloam, J. (2008) Teaching democracy: The role of political science education, *The British Journal of Politics and International Relations* 10 (3): 509–524.

Smagorinsky, P. and Taxel, J. (2005) *The Discourse of Character Education: Culture Wars in the Classroom*. New York, NY: Routledge.

Smith, A. (2003) Citizenship education in Northern Ireland: Beyond national identity? *Cambridge Journal of Education* 33 (1): 15–32.

Smith, D. M. (2000) *Moral Geographies: Ethics in a World of Difference*. Edinburgh: Edinburgh University Press?

Smith, D. P. and Mills, S. (2019) The 'youth-fullness' of youth geographies: 'coming of age'? *Children's Geographies* 17 (1): 1–8.

Smith, H. J. (2016) Britishness as racist nativism: A case of the unnamed 'other', *Journal of Education for Teaching* 42 (3): 298–313.

Spohrer, K. and Bailey, P. L. J. (2020) Character and resilience in English education policy: Social mobility, self-governance and biopolitics, *Critical Studies in Education* 61 (5): 561–576.

Springhall, J. (1977) *Youth, Empire and Society: British Youth Movements 1908–1930*. Beckenham: Croom Helm.

Springhall, J., Fraser, B. and Hoare, M. (1983) *Sure and Stedfast: A History of the Boys' Brigade 1883–1983*. London: Collins.

Staeheli, L. A. (2011) Political geography: Where's citizenship? *Progress in Human Geography* 35 (3): 393–400.

Staeheli, L. A. (2018) Learning to be citizens: Exploring social cohesion and security in times of uncertainty, *Geografiska Annaler: Series B, Human Geography* 100 (2): 61–63.

Staeheli, L. A., Attoh, K. and Mitchell, D. (2013) Contested engagements: Youth and the politics of citizenship, *Space and Polity* 17 (1): 88–105.

Staeheli, L. A. and Hammett, D. (2010) Educating the new national citizen: Education, political subjectivity and divided societies, *Citizenship Studies* 14 (6): 667–680.

Starkey, H. (2008) Diversity and citizenship in the curriculum, *London Review of Education* 6: 5–10.

Starkey, H. (2018) Fundamental British values and citizenship education: Tensions between national and global perspectives, *Geografiska Annaler: Series B, Human Geography* 100 (2): 149–162.

Stewart, H. (2020) Boris Johnson blames 'mutant algorithm' for exams fiasco, *The Guardian* 26 August 2020. Available from: https://www.theguardian.com/politics/2020/aug/26/boris-johnson-blames-mutant-algorithm-for-exams-fiasco.

Stewart, W. A. C. (1972) *Progressives and Radicals in English Education 1750–1970*. Basingstoke: Palgrave Macmillan.

Suárez, D. F. (2008) Rewriting citizenship? Civic education in Costa Rica and Argentina, *Comparative Education* 44 (4): 485–503.

Suissa, J. (2015) Character education and the disappearance of the political, *Ethics and Education* 10 (1): 105–117.

Susman, W. I. (1984) *Culture as History: The Transformation of American Society in the Twentieth Century*. New York, NY: Pantheon Books.

Sutton Trust and Social Mobility Commission (2019) *Elitist Britain 2019: The Educational Backgrounds of Britain's Leading People*. London: Sutton Trust and Social Mobility Commission.

Tan, C. and Tan, C. S. (2014) Fostering social cohesion and cultural sustainability: Character and citizenship education in Singapore, *Diaspora, Indigenous, and Minority Education* 8 (4): 191–206.

TASS (2020) Russian schools to launch student character development program starting September 1, *TASS*, 1 September 2020. Available from: https://tass.com/society/1195847.

Taylor, N. (2018) The return of character: Parallels between late-Victorian and twenty-first century discourses, *Sociological Research Online* 23 (2): 399–415.

The Jubilee Centre (2014) *Centre Publishes Statement on Youth Social Action and Character Development*, 17 November 2014. Available from: https://www.jubileecentre.ac.uk/media/news/article/4193/Centre-Publishes-Statement-on-Youth-Social-Action-and-Character-Development. Birmingham: The Jubilee Centre for Character and Virtues.

The Jubilee Centre (2015) *Polling*. Available from: https://www.jubileecentre.ac.uk/1633/character-education/resources/polling. Birmingham: The Jubilee Centre for Character and Virtues.

The Jubilee Centre and Step up to Serve (2014) *Statement on Youth Social Action and Character Development*, September 2014. Available from: https://www.jubileecentre.ac.uk/userfiles/jubileecentre/pdf/StatementSocialAction.pdf.

The Scouts (2019a) Scouts reveal six ways to build a resilient younger generation. Available from: https://www.scouts.org.uk/news/2019/december/scouts-reveals-six-ways-to-build-a-resilient-younger-generation/.

The Scouts (2019b) So, what are four and five year olds getting up to? Available from: https://www.scouts.org.uk/news/2019/december/early-years-pilot-programme/.

The Telegraph (2010) Conservatives plan civilian 'national service' scheme, *The Telegraph*. Available from: https://www.telegraph.co.uk/news/election-2010/7565987/Conservatives-plan-civilian-national-service-scheme.html.

Thomson, A. (2021) *How Covid kids can become Generation Grit*, 20 January 2021. Available from: https://www.thetimes.co.uk/article/how-covid-kids-can-become-generation-grit-0kwllxq53?-xx-meta=denied_for_visit%3D0%26visit_number%3D0%26visit_remaining%3D0%26visit_used%3D0&-xx-mvt-opted-out=false&-xx-uuid=3a49c675f71ae84064b9b230b3341376&ni-statuscode=acsaz-307.

Tyler, I. (2020) *Stigma: The Machinery of Inequality*. London: Zed Books.

Uprichard, E. (2008) Children as 'Being and Becomings': Children, childhood and temporality, *Children & Society* 22 (4): 303–313.

UNITE (2014) *The Future of Youth Work*. Available from: http://b.3cdn.net/unitevol/f9b7f490fdc5efe8c4_2rm6b3l63.pdf.

Valentine, G. (1996) Angels and devils: Moral landscapes of childhood, *Environment and Planning D: Society and Space* 14 (5): 581–599.

Valentine, G. (2003) Boundary crossings: Transitions from childhood to adulthood, *Children's Geographies* 1 (1): 37–52.

Valins, O. (2002) Defending identities or segregating communities? Faith-based schooling and the UK Jewish community, *Geoforum* 34: 235–247.

Vaughan, R. (2018) Exclusive: Damian Hinds tells schools to 'get kids climbing trees' to build character, *The Independent*. Available from: https://inews.co.uk/news/education/damian-hinds-schools-kids-climb-trees-build-character/.

Virtue Insight Blog, Birmingham: University of Birmingham. Available from: https://virtueinsight.wordpress.com/2016/06/15/is-grit-the-magic-elixir-of-good-character/.

Walker, C. (2020) Uneven solidarity: The school strikes for climate in global and inter-generational perspective, *Sustainable Earth* 3 (5): 1–13.

Walker, M., Sims, D. and Kettlewell, K. (2017) *Leading Character Education in Schools: Case Study Report*. Slough, UK: National Foundation for Educational Research. ISBN 978-1-911039-56-3.

Wall Street Journal (2020) *Kipp Wokes Up*, 6 July 2020. Available from: https://www.wsj.com/articles/kipp-wokes-up-11594076432.

Warikoo, N. K. (2016) *The Diversity Bargain: And Other Dilemmas of Race, Admissions and Meritocracy at Elite Universities*. Chicago, IL: University of Chicago Press.

Warren, A. (2004) Powell, Robert Stephenson Smyth Baden-, First Baron Baden-Powell (1857–1941), in *Oxford Dictionary of National Biography*. Cary, NC: Oxford University Press.

Webb, A. and Radcliffe, S. (2015) Indigenous citizens in the making: Civic belonging and racialized schooling in Chile, *Space and Polity* 19 (3): 215–230.

Weinberg, J. (2019) Who's listening to whom? The UK House of Lords and evidence-based policy-making on citizenship education, *Journal of Education Policy*, https://doi.org/10.1080/02680939.2019.1648877.

Weinberg, J. and Flinders, M. (2018) Learning for democracy: The politics and practice of citizenship education, *British Educational Research Journal* 44 (4): 573–592.

Weller, S. (2007) *Teenagers' Citizenship: Experiences and Education*. London: Routledge.

Wells, K. (2014) Marching to be somebody: A governmentality analysis of online cadet recruitment, *Children's Geographies* 12 (3): 339–353.

Welsh Government (2019) *A Guide to Curriculum for Wales 2022*. Available from: https://hwb.gov.wales/draft-curriculum-for-wales-2022/a-guide-to-curriculum-for-wales-2022/.

White, B. W. (1996) Talk about school: Education and the colonial project in French and British Africa (1860-1960), *Comparative Education* 32 (1): 9–26.

Whitehead, M., Lilley, R., Howell, R., Jones, R. and Pykett, J. (2016) (Re)inhabiting awareness: Geography and mindfulness, *Social & Cultural Geography* 17 (4): 553–573.

Whittaker, F. (2016) £2m 'character education' grant goes to military-style projects, *Schools Week*. Available from: https://schoolsweek.co.uk/2m-more-earmarked-for-military-style-projects/

Whittaker, F. (2018) Pupils should ditch gadgets and climb trees in 2019, Damian Hinds tells school, *Schools Week*. Available from: https://schoolsweek.co.uk/pupils-should-ditch-gadgets-and-climb-trees-in-2019-damian-hinds-tells-schools/.

Whittaker, F. (2019a) Nicky Morgan calls for return of character education grants, *Schools Week*. Available from: https://schoolsweek.co.uk/nicky-morgan-calls-for-return-of-character-education-grants/.

Whittaker, F. (2019b) Revealed: The DfE's six new character education benchmarks, *Schools Week*. Available from: https://schoolsweek.co.uk/revealed-the-dfes-six-new-character-education-benchmarks/.

Whittaker, F. and Murray, C. (2019) DfE plans new Gatsby-style 'benchmarks' for character education, *Schools Week*. Available from: https://schoolsweek.co.uk/dfe-plans-new-gatsby-style-benchmarks-for-character-education/.

Wilkinson, E. and Ortega Alcazar, I. (2017) A home of one's own? Housing welfare for 'young adults' in times of austerity, *Critical Social Policy* 37 (3): 329–347.

Williamson, B. (2019) Brain data: Scanning, scraping and sculpting the plastic learning brain through neurotechnology, *Postdigital Science and Education* 1: 65–86.

Williamson, B. (2020) Datafication of Education: A Critical Approach to Emerging Analytics Technologies and Practices, in H. Beetham and R. Sharpe (eds) *Rethinking Pedagogy for a Digital Age: Principles and Practices of Design*. London: Routledge: 212–226.

Williamson, B. (2021) Psychodata: Disassembling the psychological, economic, and statistical infrastructure of 'social-emotional learning', *Journal of Education Policy* 36 (1): 129–154.

Williamson, H. (1993) Youth policy in the United Kingdom and the marginalisation of young people, *Youth and Policy* 40: 33–48.

Winton, S. (2008) The appeal(s) of character education in threatening times: Caring and critical democratic responses, *Comparative Education* 44 (3): 305–316.

Wood, B.E. (2012) Scales of active citizenship: New Zealandzealand teachers' diverse perceptions and practices, *International Journal of Progressive Education*. 8 (3): 77–93.

Wood, J. (2010) 'Preferred Futures': Active citizenship, government and young people's voices, *Youth and Policy* 105, 50–70.

Woods, P. A., Woods, G. J. and Gunter, H. (2007) Academy schools and entrepreneurialism in education, *Journal of Education Policy* 22 (2): 237–259.

Woodyer, T. and Carter, S. (2020) Domesticating the geopolitical: Rethinking popular geopolitics through play, *Geopolitics* 25 (5): 1050–1074.

Worth, N. (2009) Understanding youth transition as 'becoming': Identity, time and futurity, *Geoforum* 40 (6): 1050–1060.

Wright, S. (2017) *Morality and Citizenship in English Schools: Secular Approaches, 1897–1944*. Basingstoke: Palgrave Macmillan.

Yang, W. (2017) Does 'compulsory volunteering' affect subsequent behaviour? Evidence from a natural experiment in Canada, *Education Economics* 25 (4): 394–405.

Yarwood, R. (2013) *Citizenship*. London: Routledge.

You, Y. and Morris, P. (2016) Imagining school autonomy in high-performing education systems: East Asia as a source of policy referencing in England, *Compare: A Journal of Comparative and International Education* 46 (6): 882–905.

Index

Printed in the United States
by Baker & Taylor Publisher Services